1-1 Reading to Learn Mathematics

A Plan for Problem Solving

AF868955

Pre-Activity ***Complete the Mini Lab at the top of page 6 in your textbook. Write your answers below.***

1. How many white tiles does it take to border each of these three gardens?

2. Predict how many white tiles it will take to border the next-largest garden. Check your answer by modeling the garden.

3. How many tiles will it take to border a garden that is 10 tiles long? Explain your reasoning.

Reading the Lesson

4. Read the Examine section in Example 1 at the bottom of page 7. In the equation $8 + 22 = 30$, what does the 8 stand for? What does the 22 stand for?

5. Look at the graph in Example 2 on page 8. What does the word "annual" mean? Now read the Plan section. Explain how to find a person's weekly salary when you know the annual salary.

6. Look again at the graph in Example 2 on page 8. Explain what the heights of the flags represent. Why do the flags have different heights?

Helping You Remember

7. *Early problem solvers excel* is a mnemonic aid to remember the first letters of the steps in the problem-solving plan. Write a mnemonic aid of your own using the first letters of the steps.

NAME ______________________________ DATE ____________ PERIOD _____

1-2 Reading to Learn Mathematics

Variables, Expressions, and Properties

Pre-Activity

Complete the Mini Lab at the top of page 11 in your textbook. Write your answers below.

1. Complete the table below.

Figure Number	1	2	3	4	5	6
Perimeter	4	8				

2. What would be the perimeter of Figure 10?

3. What is the relationship between the figure number and the perimeter of the figure?

Reading the Lesson

4. Number the operations in the correct order for simplifying $2 + 4(9 - 6 \div 3)$. Then simplify the expression.

____ Addition ____ Multiplication

____ Subtraction ____ Division

For Exercises 5–9, describe how each pair of numerical expressions is different. Then determine whether the two expressions are equal to each other. If the expressions are equal, name the property that says they are equal.

5. $2 + 5, 5 + 2$

6. $(6 - 4) - 1, 6 - (4 - 1)$

7. $2(5 - 3), 2 \cdot 5 - 2 \cdot 3$

8. $5 \cdot (4 \cdot 7), (5 \cdot 4) \cdot 7$

9. $10 \div 2, 2 \div 10$

Helping You Remember

10. The word *counter* has several meanings in the English language. Use a dictionary to find the meaning of *counter* when it is used as a prefix in the word *counterexample*. Then write your own definition of *counterexample*.

CONTENTS

Chapter 10

Chapter 11

Chapter 12

CHAPTER 1

NAME ______________________________ DATE ____________ PERIOD _____

Reading to Learn Mathematics

Vocabulary Builder

This is an alphabetical list of new vocabulary terms you will learn in Chapter 1. As you study the chapter, complete each term's definition or description. Remember to add the page number where you found the term. Add this page to your math study notebook to review vocabulary at the end of the chapter.

Vocabulary Term	Found on Page	Definition/Description/Example
absolute value		
additive inverse		
algebraic [AL-juh-BRAY-ihk] expression		
conjecture		
coordinate		
counterexample		
defining a variable		
equation [ih-KWAY-zhuhn]		
evaluate		
inequality		
integer [IHN-tih-juhr]		

NAME ______________________ DATE ____________ PERIOD _____

Reading to Learn Mathematics

Vocabulary Builder *(continued)*

Vocabulary Term	Found on Page	Definition/Description/Example
inverse operations		
negative number		
numerical expression		
open sentence		
opposites		
order of operations		
powers		
property		
solution		
solve		
variable		

NAME ______________________ DATE ____________ PERIOD _____

1-3 Reading to Learn Mathematics

Integers and Absolute Value

Pre-Activity ***Read the introduction at the top of page 17 in your textbook. Write your answers below.***

1. What does an elevation of -86 meters represent?

2. What does a temperature of $-35°$ represent?

Reading the Lesson

The symbol ... is called an ellipsis.

3. Look on page 17 in your textbook to find the meaning of the ellipsis as it is used in the list 1, 4, 7, 10,... .

4. Use a dictionary to find the meaning of the ellipsis as it is used in the sentence *The marathon began...downtown.*

5. How can you explain the usage of the ellipsis in the list in Exercise 3 in terms of the meaning for the ellipsis in the sentence in Exercise 4?

6. Look at the number line on page 17 of your textbook. How are the ellipses (plural of ellipsis) in the set of integers {..., -3, -2, -1, 0, 1, 2, 3,...} represented on the number line?

Complete each sentence with either *left* or *right* to make a true sentence. Then write a statement comparing the two numbers with either < or >.

7. -45 lies to the ________ of 0 on a number line.

8. 72 lies to the ________ of 0 on a number line.

9. -3 lies to the ________ of -95 on a number line.

10. 6 lies to the ________ of -7 on a number line.

11. Describe the symbol for the absolute value of 3. Then write the symbol.

Helping You Remember

12. Write a mathematical expression that represents the following sentence. (*Hint*: Let f represent the 49ers' Score and s represent the Seahawks' score.)
The Seahawks and the 49ers scored within 3 points of each other.

NAME ______________________ DATE ____________ PERIOD _____

1-4 Reading to Learn Mathematics

Adding Integers

Pre-Activity ***Read the introduction at the top of page 23 in your textbook. Write your answers below.***

1. Write an integer that describes the game show host's statement.

2. Write an addition sentence that describes this situation.

Reading the Lesson

3. Look at your answer for Exercise 2. Identify each number in the addition sentence as either an addend or a sum.

Identify the number with the greater absolute value.

4. 4, 8

5. −3, 5

6. 9, −12

7. −23, −16

Determine whether you *add* or *subtract* the absolute values of the numbers to find the sum. Give a reason for your answer.

8. 4 + 8

9. −3 + 5

10. 9 + (−12)

11. −23 + (−16)

Determine whether the sum is *positive* or *negative*. Then find the sum.

12. 4 + 8

13. −3 + 5

14. 9 + (−12)

15. −23 + (−16)

Add.

16. 3 + (−4)

17. −3 + 4

18. −6 + (−4)

19. 7 + 8

20. 25 + (−17)

21. −34 + (−17)

22. −43 + 4

23. 11 + (−30)

24. −81 + (−63)

25. −39 + 124

26. 97 + (−165)

27. −49 + (−75)

Helping You Remember

28. You have seen what a negative number means in terms of weather or money. Describe what a negative number means on a video cassette recorder.

NAME ______________________ DATE ____________ PERIOD ______

1-5 Reading to Learn Mathematics

Subtracting Integers

Pre-Activity ***Complete the Mini Lab at the top of page 28 in your textbook. Write your answers below.***

1. How does this result compare with the result of $3 + (-5)$?

2. Use counters to find $-4 - 2$.

3. How does this result compare to $-4 + (-2)$?

4. Use counters to find each difference and sum. Compare the results in each group.

a. $1 - 5$; $1 + (-5)$ **b.** $-6 - 4$; $-6 + (-4)$

Reading the Lesson

5. Find the opposite of 7.

6. Find the additive inverse of 7.

7. How is the opposite of a number different from the additive inverse of the number?

Rewrite each difference as a sum. Then find the sum.

8. $2 - 9$

9. $-3 - 8$

10. $10 - (-12)$

11. $-5 - (-16)$

Subtract.

12. $3 - (-5)$

13. $-3 - 5$

14. $-7 - (-3)$

15. $6 - 8$

16. $23 - (-17)$

17. $-24 - (-12)$

18. $-41 - 4$

19. $31 - (-26)$

20. $-81 - (-33)$

21. $-139 - 134$

22. $97 - (-265)$

23. $-59 - (-77)$

24. Describe the method for subtracting integers.

Helping You Remember

25. Subtraction and addition are often referred to as opposite operations. Explain in your own words the relationship between addition and subtraction.

NAME ______________________________ DATE ____________ PERIOD _____

1-6 Reading to Learn Mathematics

Multiplying and Dividing Integers

Pre-Activity ***Read the introduction at the top of page 34 in your textbook. Write your answers below.***

1. Write two different addition sentences that could be used to find the submersible's depth after 3 minutes. Then find their sum.

2. Write a multiplication sentence that could be used to find this same depth. Explain your reasoning.

3. Write a multiplication sentence that could be used to find the submersible's depth after 10 minutes. Then find the product.

Reading the Lesson

4. Identify each number in the multiplication sentence $3(-120) = -360$ as either a factor or a product.

Complete each sentence with either *positive* or *negative*.

5. The product of two integers with different signs is __________ .

6. The product of two integers with the same signs is __________ .

7. The quotient of two integers with different signs is __________ .

8. The quotient of two integers with the same signs is __________ .

Determine whether each product or quotient is *positive* or *negative*. Then evaluate the expression.

9. $4 \cdot 8$

10. $-3 \cdot 5$

11. $9(-2)$

12. $-6(-7)$

13. $12 \div (-4)$

14. $-35 \div (-7)$

15. $\frac{21}{3}$

16. $\frac{-64}{8}$

Helping You Remember

17. Explain how to find the mean of a set of numbers. What is another name for the mean?

NAME ______________________________ DATE ____________ PERIOD _____

1-7 Reading to Learn Mathematics

Writing Expressions and Equations

Pre-Activity ***Read the introduction at the top of page 39 in your textbook. Write your answers below.***

1. What is the relationship between the number of guests and the cost of the party?

2. Write an expression representing the cost of a party with *g* guests.

Reading the Lesson

Look at the table of common words and phrases on page 39. Then determine whether each situation requires *addition, subtraction, multiplication,* or *division.*

3. Find the difference between the cost of a gallon of premium gasoline and the cost of a gallon of regular gasoline.

4. Find the cost per person when the price of a pizza is split among several people.

5. Find the price of an airline ticket after the price has been decreased by $50.

6. Find how much an executive spent on breakfast, lunch, and dinner.

7. Find the flight time after the time has been increased by 15 minutes.

8. Find the product of the price of a calculator and the number of students in the class.

9. Find the high temperature on Wednesday if this temperature is 3° less than the high temperature on Tuesday.

10. Find the ratio of the amount of gasoline used and the distance traveled.

Helping You Remember

11. Devise your own way to determine whether a verbal description should be translated as an algebraic expression or an algebraic equation.

NAME ______________________ DATE ____________ PERIOD _____

1-8 Reading to Learn Mathematics

Solving Addition and Subtraction Equations

Pre-Activity ***Complete the Mini Lab at the top of page 45 in your textbook. Write your answers below.***

Solve each equation using cups and counters.

1. $x + 1 = 4$ **2.** $x + 3 = 7$ **3.** $x + (-4) = -5$

4. Explain how you would find a value of x that makes $x + (-3) = -8$ true without using models.

Reading the Lesson

5. Match the method of solving with the appropriate equation.

$x + 6 = 9$ _____ **a.** Subtract 11 from each side.

$s - 5 = 14$ _____ **b.** Subtract 6 from each side.

$4 = -3 + p$ _____ **c.** Add 3 to each side.

$11 + m = 33$ _____ **d.** Add 5 to each side.

For Exercises 6–8, explain how to solve each equation.

6. $w - 7 = -2$ ______________________

7. $c + 3 = 9$ ______________________

8. $17 = 11 + k$ ______________________

Solve each equation.

9. $z + 8 = 2$ **10.** $3 = -7 + r$ **11.** $-9 = g - 14$

Helping You Remember

12. Write two addition and two subtraction equations of your own. Trade your equations with a partner and solve. Explain to each other the method you used to solve the equations.

NAME ______________________ DATE ____________ PERIOD _____

1-9 Reading to Learn Mathematics

Solving Multiplication and Division Equations

Pre-Activity ***Read the introduction at the top of page 50 in your textbook. Write your answers below.***

1. If d represents the number of days the bamboo has been growing, write a multiplication equation you could use to find how long it would take for the bamboo to reach a height of 210 inches.

Reading the Lesson

Complete each sentence.

2. To solve $3x = 51$, __________ each side by 3.

3. To solve $\frac{b}{-2} = 4$, __________ each side by -2.

4. To solve $-65 = -5t$, __________ each side by -5.

5. To solve $-7 = \frac{d}{6}$, __________ each side by 6.

Explain how to solve each equation.

6. $\frac{u}{6} = 13$ ____________________

7. $-2c = -14$ ____________________

8. $64 = 16k$ ____________________

Solve each equation.

9. $8r = 32$

10. $3 = \frac{x}{7}$

11. $-9 = -9g$

Helping You Remember

12. Write two multiplication and two division equations of your own. Trade your equations with a partner and solve. Explain to each other the method you used to solve the equations.

NAME ______________________ DATE ____________ PERIOD _____

Reading to Learn Mathematics

Vocabulary Builder

This is an alphabetical list of new vocabulary terms you will learn in Chapter 2. As you study the chapter, complete each term's definition or description. Remember to add the page number where you found the term. Add this page to your math study notebook to review vocabulary at the end of the chapter.

Vocabulary Term	Found on Page	Definition/Description/Example
bar notation		
base		
dimensional analysis		
exponent		
like fractions		
multiplicative inverses		

NAME ______________________ DATE ____________ PERIOD _____

Reading to Learn Mathematics

Vocabulary Builder *(continued)*

Vocabulary Term	Found on Page	Definition/Description/Example
power		
rational number		
reciprocals		
repeating decimal		
scientific notation		
terminating decimal		
unlike fractions		

NAME ______________________________ DATE ____________ PERIOD _____

2-1 Reading to Learn Mathematics

Fractions and Decimals

Pre-Activity ***Read the introduction at the top of page 62 in your textbook. Write your answers below.***

1. What fraction of the sites are in the United States?

2. What fraction of the sites are in Canada?

3. At what fraction of the sites might you see gray whales?

4. What fraction of the humpback viewing sites are in Mexico?

Reading the Lesson

5. Explain the difference in meaning between the expressions $4\frac{3}{4}$ and $4\left(\frac{3}{4}\right)$.

6. Explain the difference between the numbers $2.5\overline{7}$ and $2.\overline{57}$.

Helping You Remember

7. Notice that the first five letters of the word *rational* is the word ratio. Explain what a ratio is. If this term is not familiar to you, look it up in the dictionary. Write a ratio and a rational number. Explain how they are related.

NAME ______________________ DATE ____________ PERIOD _____

2-2 Reading to Learn Mathematics

Comparing and Ordering Rational Numbers

Pre-Activity ***Read the introduction at the top of page 67 in your textbook. Write your answers below.***

1. Do we recycle more or less than half of the paper we produce? Explain.

2. Do we recycle more or less than half of the aluminum cans? Explain.

3. Which items have a recycle rate less than one half?

4. Which items have a recycle rate greater than one half?

5. Using this estimation method, can you order the rates from least to greatest?

Reading the Lesson

6. Read Example 2 on page 68. Explain how to use a number line to determine which of two rational numbers is the lesser number.

For Exercises 7 and 8, graph each pair of rational numbers on a number line. Then identify the lesser number.

7. $\frac{1}{5}, \frac{1}{3}$

8. $-\frac{4}{5}, -\frac{9}{10}$

Helping You Remember

9. Order the numbers $\frac{3}{7}, \frac{3}{5}, \frac{3}{8}, \frac{3}{4}$, and $\frac{3}{11}$ from least to greatest. Then write a rule that helps you compare two positive fractions with the same numerator.

NAME ______________________________ DATE ____________ PERIOD _____

2-3 Reading to Learn Mathematics

Multiplying Rational Numbers

Pre-Activity ***Complete the Mini Lab at the top of page 71 in your textbook. Write your answers below.***

1. What is the product of $\frac{1}{3}$ and $\frac{2}{5}$?

2. Use an area model to find each product.

a. $\frac{3}{4} \cdot \frac{1}{2}$ **b.** $\frac{2}{5} \cdot \frac{2}{3}$ **c.** $\frac{1}{4} \cdot \frac{3}{5}$ **d.** $\frac{2}{3} \cdot \frac{4}{5}$

3. What is the relationship between the numerators of the factors and the numerator of the product?

4. What is the relationship between the denominators of the factors and the denominator of the product?

Reading the Lesson

5. What is the greatest common factor of two numbers?

6. How is the greatest common factor used when multiplying fractions?

7. How is *dimensional analysis* defined on page 73 in your textbook?

8. How is dimensional analysis used in Example 5 on page 73 in your textbook?

9. Look at the table in Example 5 again. Answer the question for a different aircraft in the table.

Helping You Remember

10. If you were to visit Europe, you may need to exchange some of your money for Euros. The *exchange rate* tells you how many dollars equals how many Euros. How would you use dimensional analysis to compute the number of Euros you would get from $50.

NAME ______________________________ DATE ____________ PERIOD _____

2-4 Reading to Learn Mathematics

Dividing Rational Numbers

Pre-Activity ***Read the introduction at the top of page 76 in your textbook. Write your answers below.***

1. Find the value of $110 \div 4$.

2. Find the value of $110 \cdot \frac{1}{4}$.

3. Compare the values of $110 \div 4$ and $110 \cdot \frac{1}{4}$.

4. What can you conclude about the relationship between dividing by 4 and multiplying by $\frac{1}{4}$?

Reading the Lesson

5. Describe the process for finding the multiplicative inverse of a mixed number.

For Exercises 6–9, write the multiplicative inverse of each mixed number.

6. $2\frac{1}{5}$ **7.** $-1\frac{3}{8}$ **8.** $3\frac{4}{7}$ **9.** $5\frac{5}{9}$

10. Explain how to divide by a fraction.

11. Look at your answers for Exercises 6 and 10 above. How do you divide a number by $2\frac{1}{5}$?

Helping You Remember

12. Look up the word *invert* in the dictionary. Draw a simple picture and then invert it. Explain how this helps you remember how to divide fractions.

NAME ______________________________ DATE ____________ PERIOD _____

2-5 Reading to Learn Mathematics

Adding and Subtracting Like Fractions

Pre-Activity ***Read the introduction at the top of page 82 in your textbook. Write your answers below.***

1. What is the sum of the whole-number parts of the amounts?

2. How many $\frac{1}{3}$ cups are there?

3. Since $\frac{1}{3} + \frac{1}{3} + \frac{1}{3} = 1$, how many cups do all the $\frac{1}{3}$ cups make?

4. What is the total number of cups of the ingredients listed?

Reading the Lesson

5. Define *like fractions*.

For Exercises 6–9, determine whether each pair of fractions are like fractions.

6. $\frac{3}{5}, \frac{3}{7}$ **7.** $\frac{5}{8}, \frac{7}{8}$ **8.** $\frac{4}{7}, -\frac{5}{7}$ **9.** $\frac{5}{9}, -\frac{2}{3}$

10. Explain how to add like fractions.

11. Explain how to subtract like fractions.

Add or subtract. Write in simplest form.

12. $\frac{3}{5} + \frac{1}{5}$ **13.** $\frac{5}{8} + \frac{7}{8}$ **14.** $\frac{5}{9} - \frac{2}{9}$ **15.** $\frac{4}{7} - \frac{5}{7}$

Helping You Remember

16. Talk with a partner about the word *like*. What does it usually mean? How is this different from the way it is used in the lesson?

NAME ______________________________ DATE ____________ PERIOD _____

2-6 Reading to Learn Mathematics

Adding and Subtracting Unlike Fractions

Pre-Activity ***Read the introduction at the top of page 88 in your textbook. Write your answers below.***

1. What are the denominators of the fractions?

2. What is the least common multiple of the denominators?

3. Find the missing value in $\frac{1}{4} = \frac{?}{8}$.

4. What fraction of the pizza did the two girls eat?

Reading the Lesson

5. What do *LCM* and *LCD* stand for? Give a definition for each.

Find the LCM of each pair of numbers.

6. 2, 3 **7.** 4, 6 **8.** 5, 10 **9.** 9, 12

Find the LCD of each pair of fractions.

10. $\frac{3}{5}, \frac{3}{7}$ **11.** $\frac{5}{8}, \frac{7}{12}$ **12.** $\frac{4}{7}, -\frac{5}{7}$ **13.** $\frac{5}{9}, -\frac{2}{3}$

14. Explain how to add or subtract unlike fractions.

Rewrite each sum or difference in terms of like fractions. Then add or subtract. Write in simplest form.

15. $\frac{3}{5} + \frac{1}{2}$ **16.** $\frac{3}{4} + \frac{7}{8}$ **17.** $\frac{5}{9} - \frac{2}{3}$

18. $\frac{4}{7} - \frac{1}{2}$ **19.** $\frac{3}{5} + \frac{3}{7}$ **20.** $\frac{5}{8} - \frac{7}{12}$

Helping You Remember

21. Describe what the prefix *un-* usually means when it appears in front of a word. How does this meaning relate to unlike fractions?

NAME ______________________ DATE ____________ PERIOD _____

2-7 Reading to Learn Mathematics

Solving Equations with Rational Numbers

Pre-Activity ***Read the introduction at the top of page 92 in your textbook. Write your answers below.***

1. Multiply each side of the equation by 6. Write the result.

2. Divide each side of the equation in Exercise 1 by 5. Write the result.

3. Multiply each side of the original equation $25 = \frac{5}{6}s$ by the multiplicative inverse of $\frac{5}{6}$. Write the result.

4. What is the speed of a grizzly bear?

5. Which method of solving the equation seems most efficient? Explain.

Reading the Lesson

6. Match the method of solving with the appropriate equation.

$1.25a = 3.75$	_____	**a.** Subtract $\frac{3}{5}$ from each side.
$x + 1.25 = 5.25$	_____	**b.** Multiply each by $\frac{5}{3}$.
$\frac{3}{5}m = \frac{7}{10}$	_____	**c.** Add 1.25 to each side.
$r - 1.25 = 4.5$	_____	**d.** Divide each side by 1.25.
$\frac{3}{5} + f = \frac{1}{2}$	_____	**e.** Subtract 1.25 from each side.

Explain in words how to solve each equation.

7. $\frac{y}{3.2} = 1.1$

8. $\frac{3}{8} + v = \frac{7}{12}$

Helping You Remember

9. The description of a problem often has more information than you need to design an equation and solve it. Describe the process of writing an equation to solve a problem.

NAME ______________________ DATE ____________ PERIOD ______

2-8 Reading to Learn Mathematics

Powers and Exponents

Pre-Activity ***Read the introduction at the top of page 98 in your textbook. Write your answers below.***

1. How many 2s are multiplied to determine the number of great grandparents?

2. How many 2s would you multiply to determine the number of great-great grandparents?

Reading the Lesson

3. Define the terms *base*, *exponent*, and *power*.

For Exercises 4–6, identify the base, exponent, and power in each expression.

4. 5^4

5. 7^{-2}

6. x^8

7. Explain in words what 5^4 means.

Rewrite each expression using multiplication instead of an exponent.

8. 5^4 **9.** 9^5 **10.** c^8

Evaluate each expression.

11. 5^4 **12.** 9^5 **13.** 6^3 **14.** 2^8

Helping You Remember

15. Notice that $4^{-3} = \frac{1}{4^3}$. A power with a negative exponent is not negative. Write a true sentence using the terms *negative exponent*, *power*, *positive*, and *rational*.

NAME ______________________ DATE ____________ PERIOD _____

2-9 Reading to Learn Mathematics

Scientific Notation

Pre-Activity ***Read the introduction at the top of page 104 in your textbook. Write your answers below.***

1. All of the values contain 10^8. What is the value of 10^8?

2. How many people speak Mandarin as their native language?

3. How many people speak English as their native language?

Reading the Lesson

4. How can you tell that a number is in standard form?

Identify each positive number as either very large or very small.

5. 9,245,000

6. 0.00083986

7. 0.0000003

8. 1,000,000,000

For each pair of numbers, determine how many places the decimal has moved and whether the exponent of the original would be *positive* or *negative* in scientific notation.

9. $0.00037 \rightarrow 3.7$

10. $185{,}000 \rightarrow 1.85$

Write each number in scientific notation.

11. 8,790,000

12. 0.0000125

13. 0.00899

14. 402,500,000

Helping You Remember

15. Work with a partner. One person should explain how to write a very large number in scientific notation. The other person should explain how to write a very small number in scientific notation.

CHAPTER 3

NAME ______________________ DATE ____________ PERIOD _____

Reading to Learn Mathematics

Vocabulary Builder

This is an alphabetical list of new vocabulary terms you will learn in Chapter 3. As you study the chapter, complete each term's definition or description. Remember to add the page number where you found the term. Add this page to your math study notebook to review vocabulary at the end of the chapter.

Vocabulary Term	Found on Page	Definition/Description/Example
abscissa [ab-SIH-suh]		
converse		
coordinate plane		
hypotenuse		
irrational number		
legs		
ordered pair		
ordinate [OR-din-it]		
origin		
perfect square		

NAME ______________________ DATE ____________ PERIOD _____

Reading to Learn Mathematics

Vocabulary Builder *(continued)*

Vocabulary Term	Found on Page	Definition/Description/Example
principal square root		
Pythagorean Theorem		
Pythagorean triple		
quadrants		
radical sign		
real number		
right triangle		
square root		
x-axis		
x-coordinate		
y-axis		
y-coordinate		

NAME ______________________________ DATE ____________ PERIOD _____

3-1 Reading to Learn Mathematics

Square Roots

Pre-Activity ***Complete the Mini Lab at the top of page 116 in your textbook. Write your answers below.***

1. Copy and complete the following table.

Tiles on a Side	1	2	3	4	5
Total Number of Tiles in the Square Arrangement	1	4			

2. Suppose a square arrangement has 36 tiles. How many tiles are on a side?

3. What is the relationship between the number of tiles on a side and the number of tiles in the arrangement?

Reading the Lesson

4. The opposite of ______________________________ is finding one of two equal factors of a number.

5. Explain how you know whether a square root is the principal square root or not.

6. To solve an equation in which one side of the equation is a squared term, what can you do to each side of the equation?

Helping You Remember

7. Given enough time and enough tiles, how might you go about determining whether a whole number is a perfect square?

NAME ______________________________ DATE ______________ PERIOD ______

3-2 Reading to Learn Mathematics

Estimating Square Roots

Pre-Activity ***Complete the Mini Lab at the top of page 120 in your textbook. Write your answers below.***

1. How many squares are on each side of the largest possible square using no more than 40 small squares?

2. How many squares are on each side of the smallest possible square using at least 40 small squares?

3. The value of $\sqrt{40}$ is between two consecutive whole numbers. What are the numbers?

Use grid paper to determine between which two consecutive whole numbers each value is located.

4. $\sqrt{23}$

5. $\sqrt{52}$

6. $\sqrt{27}$

7. $\sqrt{18}$

Reading the Lesson

8. Explain how you can estimate the square root of a number if you know perfect squares greater than and less than the number.

For Exercises 9–12, estimate to the nearest whole number.

9. $\sqrt{33}$

10. $\sqrt{71}$

11. $\sqrt{114}$

12. $\sqrt{211}$

13. Read Example 2 on page 121 of your textbook. What is a "golden rectangle"?

Helping You Remember

14. Draw a triangle and label its sides. (Make sure your triangle is a real triangle. For example, sides of lengths 2, 2 and 8 do not make a triangle.) Trade triangles with a partner and estimate the area of your triangles using Heron's Formula.

NAME ______________________ DATE ____________ PERIOD _____

3-3 Reading to Learn Mathematics

The Real Number System

Pre-Activity ***Read the introduction at the top of page 125 in your textbook. Write your answers below.***

1. The length of the court is 60 feet. Is this number a whole number? Is it a rational number? Explain.

2. The distance from the net to the rear spikers line is $7\frac{1}{2}$ feet. Is this number a whole number? Is it a rational number? Explain.

3. The diagonal across the court is $\sqrt{4,500}$ feet. Can this square root be written as a whole number? a rational number?

Reading the Lesson

4. What do rational and irrational numbers have in common? What is the difference between rational numbers and irrational numbers? Give an example of each.

5. Match the property of real numbers with the algebraic example.

Commutative	_____	**a.** $(x + y) + z = x + (y + z)$
Associative	_____	**b.** $pq = qp$
Distributive	_____	**c.** $h + 0 = h$
Identity	_____	**d.** $c + (-c) = 0$
Inverse	_____	**e.** $x(y + z) = xy + xz$

Helping You Remember

6. Think of a way to remember the relationships between the sets of numbers in the real number system. For example, think of a rhyme that tells the order of the sets of numbers from smallest to largest.

3-4 Reading to Learn Mathematics

The Pythagorean Theorem

Pre-Activity ***Complete the Mini Lab at the top of page 132 in your textbook. Write your answers below.***

1. Find the area of each square.

2. How are the squares of the sides related to the areas of the squares?

3. Find the sum of the areas of the two smaller squares. How does the sum compare to the area of the larger square?

4. Use grid paper to cut out three squares with sides 5, 12, and 13 units. Form a triangle with these squares. Compare the sum of the areas of the two smaller squares with the area of the larger square.

Reading the Lesson

5. Is it possible to have a right triangle for which the Pythagorean Theorem is *not* true?

6. If you know the lengths of two of the sides of a right triangle, how can you find the length of the third side?

Use the Pythagorean Theorem to determine whether each of the following measures of the sides of a triangle are the sides of a right triangle.

7. 4, 5, 6

8. 9, 12, 15

9. 10, 24, 26

10. 5, 7, 9

Helping You Remember

11. In everyday language, a *leg* is a limb used to support the body. How does this meaning relate to the legs of a right triangle?

3-5 Reading to Learn Mathematics

Using the Pythagorean Theorem

Pre-Activity ***Read the introduction at the top of page 137 in your textbook. Write your answers below.***

1. What type of triangle is formed by the sides of the mat and the diagonal?

2. Write an equation that can be used to find the length of the diagonal.

Reading the Lesson

Determine whether each of the following is a Pythagorean triple.

3. 13-84-85

4. 11-60-61

5. 21-23-29

6. 12-25-37

7. The triple 8-15-17 is a Pythagorean triple. Complete the table to find more Pythagorean triples.

	a	b	c	Check: $c^2 = a^2 + b^2$
original	8	15	17	$289 = 64 + 225$
$\times 2$				
$\times 3$				
$\times 5$				
$\times 10$				

8. If the sides of a square are of length s, how can you find the length of a diagonal of the square?

Helping You Remember

9. Work with a partner. Write a word problem that can be solved using the Pythagorean Theorem, including the art. Exchange problems with your partner and solve.

3-6 Reading to Learn Mathematics

Distance on the Coordinate Plane

Pre-Activity

Read the introduction at the top of page 142 in your textbook. Write your answers below.

1. What type of triangle is formed by the blue and red lines?

2. What is the length of the two red lines?

3. Write an equation you could use to determine the distance d between the locations where the ring and necklace were found.

4. How far apart were the ring and the necklace?

Reading the Lesson

5. On the coordinate plane, what are the four sections determined by the axes called?

6. Match each term of the coordinate plane with its description.

ordinate	_____	**a.** point where number lines meet
y-axis	_____	**b.** x-coordinate
origin	_____	**c.** y-coordinate
abscissa	_____	**d.** vertical number line
x-axis	_____	**e.** horizontal number line

7. To find the distance between two points, draw a right triangle whose hypotenuse is the distance you want to find; find the lengths of the legs, and use ______________________________ to solve the problem.

Helping You Remember

8. Think of a way to remember the names of the four quadrants of the coordinate plane.

NAME ______________________ DATE ____________ PERIOD _____

Reading to Learn Mathematics

Vocabulary Builder

This is an alphabetical list of new vocabulary terms you will learn in Chapter 4. As you study the chapter, complete each term's definition or description. Remember to add the page number where you found the term. Add this page to your math study notebook to review vocabulary at the end of the chapter.

Vocabulary Term	Found on Page	Definition/Description/Example
congruent		
corresponding parts		
cross products		
dilation [deye-LAY-shuhn]		
indirect measurement		
polygon		
proportion		
rate		
rate of change		

NAME ______________________________ DATE ____________ PERIOD _____

Reading to Learn Mathematics

Vocabulary Builder *(continued)*

Vocabulary Term	Found on Page	Definition/Description/Example
ratio		
rise		
run		
scale		
scale drawing		
scale factor		
scale model		
similar		
slope		
unit rate		

4-1 Reading to Learn Mathematics

Ratios and Rates

Pre-Activity ***Read the introduction at the top of page 156 in your textbook. Write your answers below.***

1. Which combination of ingredients would you use to make a smaller amount of the same recipe? Explain.

2. In order to make the same recipe of trail mix, how many scoops of peanuts should you use for every scoop of raisins?

Reading the Lesson

3. What does it mean if the ratio of red marbles to blue marbles is 3 to 5?

4. What is another way to write the ratio *3 to 5*?

5. What must you do before you can simplify the ratio *30 minutes to 8 hours*? What is the simplified ratio?

Helping You Remember

6. When you go to a bank to exchange money of one currency for another, the bank uses a conversion rate to calculate the amount of money in the new currency. Find out what the current conversion rate is to exchange U.S. dollars to Canadian dollars at a local bank. Then write the rate as a ratio of one currency compared to the other.

NAME ______________________________ DATE ____________ PERIOD _____

4-2 Reading to Learn Mathematics

Rate of Change

Pre-Activity ***Read the introduction at the top of page 160 in your textbook. Write your answers below.***

1. By how many bears did Alicia's collection increase between 1997 and 1999? Between 1999 and 2002?

2. Between which years did Alicia's collection increase the fastest?

Reading the Lesson

3. What does a rate of change measure on a graph?

4. On a graph, what does it mean when a rate of change is negative?

5. Complete the sentence: When a quantity does not change over a period of time, it is said to have a __________ rate of change.

Helping You Remember

6. Write out in words the formula for finding a rate of change between two data points (x_1, y_1) and (x_2, y_2).

4-3 Reading to Learn Mathematics

Slope

Pre-Activity ***Read the introduction at the top of page 166 in your textbook. Write your answers below.***

1. Pick several pairs of points from those plotted and find the rate of change between them. Write each rate in simplest form.

2. What is true of these rates?

Reading the Lesson

3. Slope can be positive, negative, or zero. For each one of these possibilities draw an example of a line with that kind of slope.

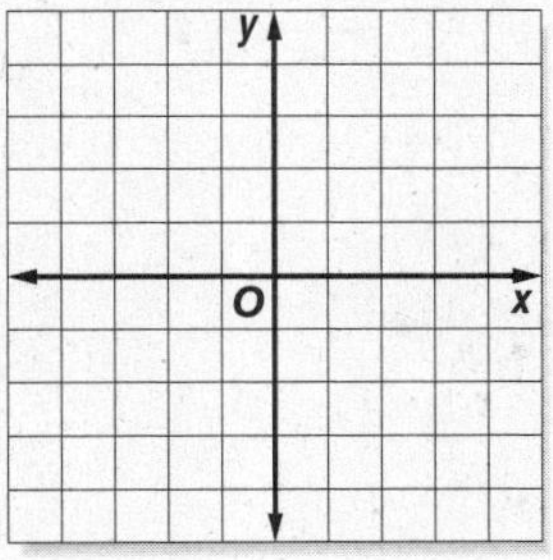

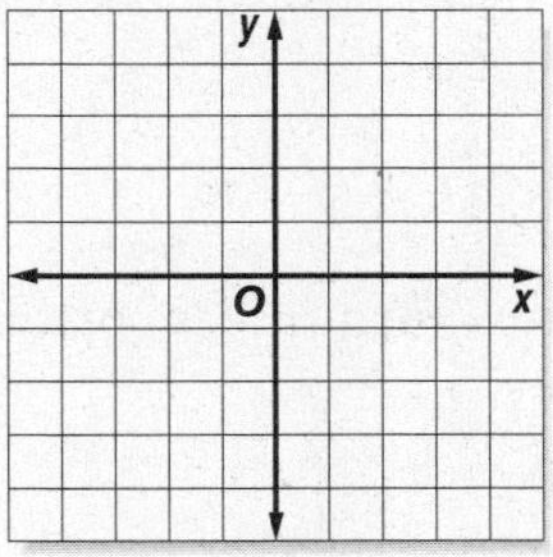

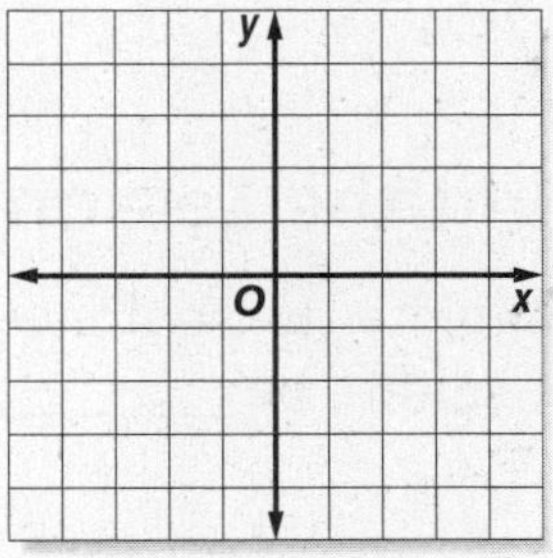

Helping You Remember

4. The slope of a line is defined by the rise compared to the run between any two points on the line. Think about the words *rise* and *run*. How can you remember which word represents vertical change and which word represents horizontal change?

NAME ______________________________ DATE ____________ PERIOD _____

4-4 Reading to Learn Mathematics

Solving Proportions

Pre-Activity ***Read the introduction at the top of page 170 in your textbook. Write your answers below.***

1. Write a ratio that compares the number of Calories from fat to the total number of Calories. Write the ratio as a fraction in simplest form.

2. Suppose you plan to eat two such granola bars. Write a ratio comparing the number of Calories from fat to the total number of Calories.

3. Is the ratio of Calories the same for two granola bars as it is for one granola bar? Why or why not?

Reading the Lesson

4. Complete the sentence: If two ratios form a proportion, then the ratios are said to be __________.

5. Do the ratios $\frac{a}{b}$ and $\frac{c}{d}$ always form a proportion? Why or why not?

6. Explain how you can use cross products to solve proportions in which one of the terms is not known.

Helping You Remember

7. For the proportion $\frac{a}{b}$ and $\frac{c}{d}$, why do you think the products ad and bc are called *cross products*?

4-5 Reading to Learn Mathematics

Similar Polygons

Pre-Activity ***Complete the Mini Lab at the top of page 178 in your textbook. Write your answers below.***

1. Compare the angles of the triangles by matching them up. Identify the angle pairs that have equal measure.

2. Express the ratios $\frac{DF}{LK}$, $\frac{EF}{JK}$, and $\frac{DE}{LJ}$ to the nearest tenth.

3. What do you notice about the ratios of the matching sides of matching triangles?

Reading the Lesson

4. Complete the sentence: If two polygons are similar, then their corresponding angles are __________, and their corresponding sides are __________.

5. If two polygons have corresponding angles that are congruent, does that mean that the polygons are similar? Why or why not?

6. If the sides of one square are 3 centimeters and the sides of another square are 9 centimeters, what is the ratio of corresponding sides from the first square to the second square?

Helping You Remember

7. Look up the everyday definition of the word *similar* in a dictionary. How does the definition relate to what you learned in this lesson?

NAME ______________________________ DATE ____________ PERIOD _____

4-6 Reading to Learn Mathematics

Scale Drawings and Models

Pre-Activity ***Read the introduction at the top of page 184 in your textbook. Write your answers below.***

1. How many units wide is the room?

2. The actual width of the room is 18 feet. Write a ratio comparing the drawing width to the actual width.

3. Simplify the ratio you found and compare it to the scale shown at the bottom of the drawing.

Reading the Lesson

4. Give another example of a scale drawing or scale model that is different from the examples of scale drawings and scale models given on pages 184–185 in your textbook.

5. Complete the sentence: distances on a scale model are __________ to distances in real life.

6. What is the scale factor for a model if part of the model that is 4 inches corresponds to a real-life object that is 16 inches?

Helping You Remember

7. Make a scale drawing of a room, such as your classroom or your bedroom. Select an appropriate scale so that your drawing is a reasonable size. Be sure to indicate your scale on your drawing. Use another piece of paper if necessary.

NAME ______________________ DATE ____________ PERIOD _____

4-7 Reading to Learn Mathematics

Indirect Measurement

Pre-Activity ***Read the introduction at the top of page 188 in your textbook. Write your answers below.***

1. How is the caveman measuring the distance to the Sun?

Reading the Lesson

2. Complete the following sentence.
When you solve a problem using shadow reckoning, the objects being compared and their shadows form two sides of ______________.

3. Suppose that you are standing near a building and you see the shadows cast by you and the building. If you know the length of each of these shadows and you know how tall you are, write a proportion in words that you can use to find the height of the building.

4. **STATUE** If a statue casts a 6-foot shadow and a 5-foot mailbox casts a 4-foot shadow, how tall is the statue?

Helping You Remember

5. Work with a partner. Have your partner draw two triangles that are similar with the lengths of two corresponding sides labeled and the length of one additional side labeled. Tell your partner how to write a proportion to solve for the length of the side corresponding to the additional side labeled.

NAME ______________________ DATE ____________ PERIOD _____

4-8 Reading to Learn Mathematics

Dilations

Pre-Activity ***Complete the Mini Lab at the top of page 194 in your textbook. Write your answers below.***

1. Multiply each coordinate by 2 to find the coordinates of points A', B', and C'.

2. On the same coordinate plane, graph points A', B', and C'. Then draw $\triangle A'B'C'$.

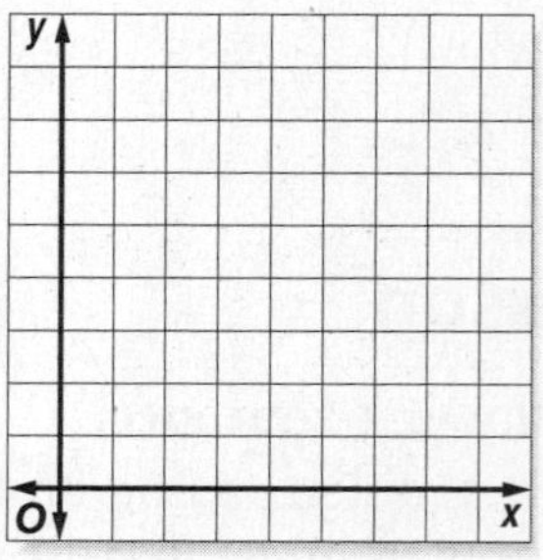

3. Determine whether $\triangle ABC \sim \triangle A'B'C'$. Explain your reasoning.

Reading the Lesson

4. If you are given the coordinates of a figure and the scale factor of a dilation of that figure, how can you find the coordinates of the new figure?

5. When you graph a figure and its image after a dilation, how can you check your work?

Helping You Remember

6. Complete the table below to help you remember the effects of different scale factors.

If the scale factor is	Then the dilation is
between 0 and 1	
greater than 1	
equal to 1	

NAME ______________________ DATE ____________ PERIOD _____

Reading to Learn Mathematics

Vocabulary Builder

This is an alphabetical list of new vocabulary terms you will learn in Chapter 5. As you study the chapter, complete each term's definition or description. Remember to add the page number where you found the term. Add this page to your math study notebook to review vocabulary at the end of the chapter.

Vocabulary Term	Found on Page	Definition/Description/Example
base		
compatible numbers		
discount		
interest		
markup		
part		
percent		

NAME ______________________________ DATE ______________ PERIOD _____

Reading to Learn Mathematics

Vocabulary Builder *(continued)*

Vocabulary Term	Found on Page	Definition/Description/Example
percent equation		
percent of change		
percent of decrease		
percent of increase		
percent proportion		
principal		
selling price		

NAME ______________________ DATE __________ PERIOD ____

5-1 Reading to Learn Mathematics

Ratios and Percents

Pre-Activity ***Read the introduction at the top of page 206 in your textbook. Write your answers below.***

1. Name two states from the table that have ratios in which the second numbers are the same.

2. Of the two states you named in Exercise 1, which state has a greater ratio of people under 18 to total population? Explain.

3. Describe how to determine which of the four states has the greater ratio of people under 18 to total population.

Reading the Lesson

4. How is *percent* defined on page 206 of your textbook?

For Exercises 5 and 6, use the grid at the right.

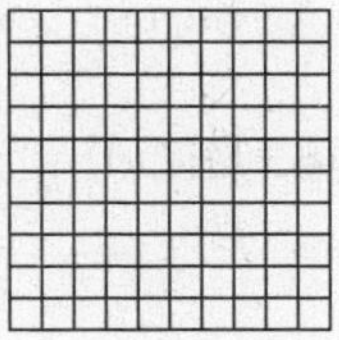

5. Describe how the grid at the right can be used to represent percents.

6. Fill in the grid to represent 55%.

7. What ratio is represented by each grid? How do the two ratios compare to each other? What percent do they represent?

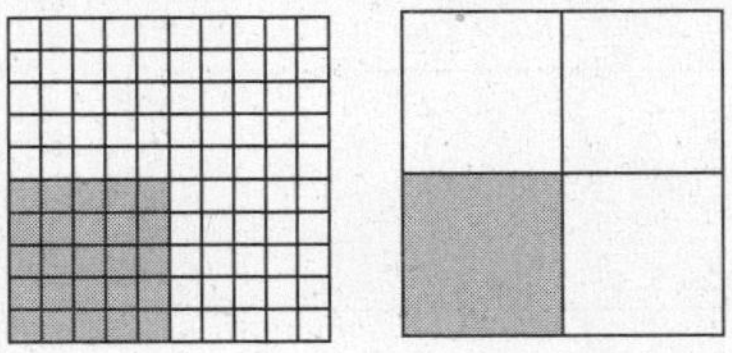

Helping You Remember

8. Work with a partner. Make up a silly statement involving a ratio, for example, "4 out of 5 of the diddlewads were goraks." Your partner writes the ratio as a fraction and a percent. Then your partner makes up a different percent and you write the percent as a fraction in simplest form. Start again with your partner making up a silly ratio statement this time.

NAME ______________________________ DATE ____________ PERIOD _____

5-2 Reading to Learn Mathematics

Fractions, Decimals, and Percents

Pre-Activity ***Read the introduction at the top of page 210 in your textbook. Write your answers below.***

1. Write each percent as a fraction. Do not simplify the fractions.

2. Write each fraction in Exercise 1 as a decimal.

3. How could you write a percent as a decimal without writing the percent as a fraction first?

Reading the Lesson

Complete each sentence.

4. To write a decimal as a percent, ____________ by 100 and add the percent symbol.

5. To write a percent as a decimal, ____________ by 100 and remove the percent symbol.

6. Dividing by 100 is the same as moving the decimal point two places to the ____________.

7. Multiplying by 100 is the same as moving the decimal point two places to the ____________.

Determine whether each expression represents changing from a decimal to a percent or a percent to a decimal.

8. 0.435 ____________________

9. 14.5 ____________________

10. 33.60 ____________________

11. 07.819 ____________________

Helping You Remember

12. Work with a partner. Pretend your partner has not studied this lesson. On a piece of paper, write a percent and a decimal using different numbers. Teach your partner how to write a percent as a decimal and how to write a decimal as a percent. Be sure to show how to place the decimal point.

NAME ______________________ DATE ____________ PERIOD _____

5-3

Reading to Learn Mathematics

The Percent Proportion

Pre-Activity ***Complete the Mini Lab at the top of page 216 in your textbook. Write your answers below.***

1. Draw a model and find the percent that is represented by each ratio.

a. 1 out of 2 **b.** 7 out of 10 **c.** 2 out of 5

2. Write a proportion that you could use to determine the percent represented by 9 out of 25.

Reading the Lesson

3. Look at page 216 in your textbook. Fill in the blanks to complete the percent proportion.

$$\frac{\_\_\_\_}{\_\_\_\_} = \frac{\_\_\_\_}{100}$$

4. Complete the table for each statement or problem. For a quantity that needs to be found, put a question mark in the appropriate column.

	part	base	percent
a. 14 is 20% of 70.			
b. 6% of 40 is 2.4			
c. 13 out of 25 is 52%			
d. What is 30% of 65?			
e. Find 41% of 250.			
f. What percent of 25 is 18?			

Helping You Remember

5. Use a clean sheet of paper and Examples 1–3 on pages 216 and 217 in your textbook. Starting with Example 1, cover up everything in the example with your paper except the title and its question. Now try to work the problem without looking at the book. Then compare your work to the work in the book. Repeat this with the other two examples.

NAME ______________________ DATE ____________ PERIOD _____

5-4 Reading to Learn Mathematics

Finding Percents Mentally

Pre-Activity ***Read the introduction at the top of page 220 in your textbook. Write your answers below.***

1. 50% of the eighth grade class are females. Write 50% as a fraction.

2. Explain how you could find 50% of 104 without using a proportion.

3. Use mental math to find the number of females in the eighth grade class.

4. 25% of the sixth grade class play intramural basketball. Write 25% as a fraction.

5. Use mental math to find the number of students in the sixth grade who play intramural basketball.

Reading the Lesson

6. Complete the following table.

Percent	Fraction	Decimal
25%		
	$\frac{1}{10}$	

Complete each statement.

7. 40% of 25 = _____ of 25 or 10

8. _____ of 36 = $\frac{1}{4}$ of 36 or 9

9. $66\frac{2}{3}\%$ of 48 = _____ of 48 or 32

10. _____ of 89 = 0.1 of 89 or 8.9

Helping You Remember

11. Work alone or with a partner. Look at the Percent-Fraction Equivalents table at the bottom of page 220 in your textbook. Create your own table on a sheet of paper or poster paper. Underneath each equivalent percent and fraction, write an example in which knowing the fraction helps you find the percent mentally.

NAME ______________________________ DATE ____________ PERIOD _____

5-5 Reading to Learn Mathematics

Percent and Estimation

Pre-Activity ***Read the introduction at the top of page 228 in your textbook. Write your answers below.***

1. Round the total area of Earth to the nearest hundred million square miles.

2. Round the percent of Earth that is land to the nearest ten percent.

3. Use what you learned about mental math in Lesson 5-4 to estimate the area of the land on Earth.

Reading the Lesson

4. What are compatible numbers?

5. Are $\frac{1}{8}$ and 56 compatible numbers? Explain.

6. Are $\frac{6}{7}$ and 32 compatible numbers? Explain.

Helping You Remember

Describe how to estimate the following using compatible numbers.

7. 65% of 64

8. 18 out of 59

NAME ______________________ DATE ____________ PERIOD ______

5-6 Reading to Learn Mathematics

The Percent Equation

Pre-Activity ***Read the introduction at the top of page 232 in your textbook. Write your answers below.***

1. Use the percent proportion to find the area of water in Florida.

2. Express the percent for Florida as a decimal. Multiply the total area of Florida by this decimal.

3. How are the answers for Exercises 1 and 2 related?

Reading the Lesson

4. What is the *percent equation*?

Write each percent proportion as a percent equation.

5. $\frac{16}{64} = \frac{25}{100}$ ______________

6. $\frac{x}{14} = \frac{2}{100}$ ______________

7. $\frac{96}{b} = \frac{48}{100}$ ______________

8. $\frac{13}{100} = \frac{p}{675}$ ______________

Helping You Remember

9. Write the percent equation in its three forms. Then choose the best form to find the total price of a jacket after sales tax. Use the sales tax percent for where you live. Find out or estimate to the nearest whole number what you think a jacket will cost where you live.

NAME ______________________________ DATE ____________ PERIOD _____

5-7 Reading to Learn Mathematics

Percent of Change

Pre-Activity *Read the introduction at the top of page 236 in your textbook. Write your answers below.*

1. How much did the price increase from 1930 to 1940?

2. Write the ratio $\frac{\text{amount of increase}}{\text{price in 1930}}$. Then write the ratio as a percent.

3. How much did the price increase from 1940 to 1950? Write the ratio $\frac{\text{amount of increase}}{\text{price in 1940}}$. Then write the ratio as a percent.

4. How much did the price increase from 1950 to 1960? Write the ratio $\frac{\text{amount of increase}}{\text{price in 1950}}$. Then write the ratio as a percent.

5. Compare the amount of increase for each ten-year period.

6. Compare the percents in Exercises 2–4.

7. Make a conjecture about why the amounts of increase are the same but the percents are different.

Reading the Lesson

8. Explain the relationship between selling price and markup.

Helping You Remember

9. When a book has many new terms or ideas, you can sometimes make an outline or concept map to help you understand the information. Read about the new terms on page 236 and the ones just before each example on pages 237 and 238. Then complete the concept map below using these words: discount, more than, new amount, markup, decrease, increase.

Percent of Change

_______ is less than the original amount	_______ is _______ the original amount
percent of _______	percent of _______
Example: _______	Example: _______

NAME ______________________ DATE ____________ PERIOD _____

5-8 Reading to Learn Mathematics

Simple Interest

Pre-Activity

Read the introduction at the top of page 241 in your textbook. Write your answers below.

1. If Hector puts his money in a savings account, he will receive 2.5% of $1,000 in interest for one year. Find the interest Hector will receive.

2. Find the interest Hector will receive if he puts his money in a money market for one year.

3. Find the interest Hector will receive if he puts his money in a certificate of deposit for one year.

Reading the Lesson

4. Look up the word *interest* in a dictionary. Read some of the different definitions. Find and write a definition that seems to match the topic in your textbook.

5. Read the paragraph just below the activity on page 241 in your textbook. What is the difference between how interest is applied to a customer for a savings account and for a credit card?

6. The words *principle* and *principal* are often confused. Look up both words in a dictionary. Which word would you use to describe the person in charge of your school and also the money invested or borrowed? Which word would you use to describe a rule of conduct?

Helping You Remember

7. Find out about what interest rate a bank near you pays on a savings account and what interest rate it charges on a credit card. Compute how much interest you would earn on $600 after 2 years. Then compute how much interest you would have to pay on a credit card if you borrowed $600 for 2 years.

CHAPTER 6

NAME ______________________________ DATE ____________ PERIOD _____

Reading to Learn Mathematics

Vocabulary Builder

This is an alphabetical list of new vocabulary terms you will learn in Chapter 6. As you study the chapter, complete each term's definition or description. Remember to add the page number where you found the term. Add this page to your math study notebook to review vocabulary at the end of the chapter.

Vocabulary Term	Found on Page	Definition/Description/Example
acute triangle		
adjacent angles		
alternate exterior angles		
alternate interior angles		
complementary angles		
corresponding angles		
equilateral triangle		
isosceles triangle		
line symmetry		
obtuse triangle		

NAME ______________________________ DATE ____________ PERIOD _____

Reading to Learn Mathematics

Vocabulary Builder *(continued)*

Vocabulary Term	Found on Page	Definition/Description/Example
parallelogram		
quadrilateral		
reflection		
rhombus		
right triangle		
rotation		
scalene triangle		
supplementary angles		
translation		
transversal		
trapezoid		
vertical angles		

6-1 Reading to Learn Mathematics

Line and Angle Relationships

Pre-Activity ***Complete the Mini Lab at the top of page 256 in your textbook. Write your answers below.***

1. For each set of intersecting lines, identify the pairs of angles that have the same measure.

2. What is true about the sum of the measures of the angles sharing a side?

Reading the Lesson

3. Angles can be acute, right, obtuse, or straight. For each one of these possibilities draw an example of the angle.

acute angle

right angle

obtuse angle

straight angle

4. How do you know if two lines are perpendicular?

5. How can you tell the difference between alternate interior and alternate exterior angles?

Helping You Remember

6. Work with a partner. Have your partner draw two angles. Identify the types of angles your partner has drawn.

NAME ______________________________ DATE ____________ PERIOD _____

6-2 Reading to Learn Mathematics

Triangles and Angles

Pre-Activity ***Complete the Mini Lab at the top of page 262 in your textbook. Write your answers below.***

1. What do you think is the sum of the measures of the three angles of any triangle? Explain your reasoning.

Reading the Lesson

2. How can you determine whether a triangle is a right triangle?

3. If one angle of a triangle is either right or obtuse, what must be true of the other two angles?

4. Complete the table.

Type of Triangle	Number of Congruent Sides
Scalene	
Isosceles	
Equilateral	

Helping You Remember

5. Complete the diagram by drawing an example of the triangle under each box.

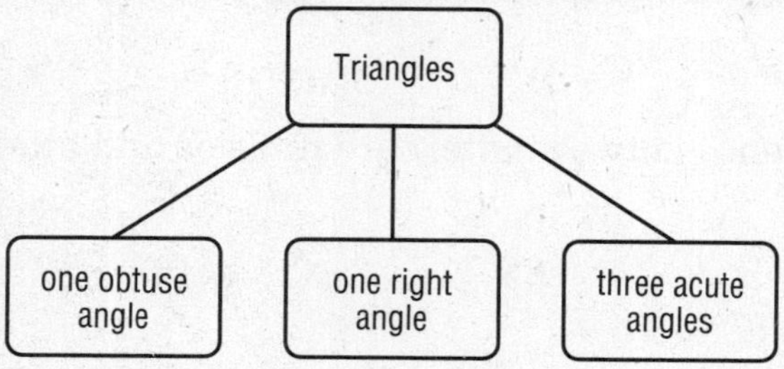

6-3 Reading to Learn Mathematics

Special Right Triangles

Pre-Activity ***Complete the Mini Lab at the top of page 267 in your textbook. Write your answers below.***

1. What type of triangles have you formed?

2. What are the measures of the angles of the folded triangle?

3. Measure and describe the relationship between the shortest and longest sides of this triangle.

4. What are the measures of the angles of the triangle formed by folding the square?

5. Measure and describe the relationship between the legs of this triangle.

Reading the Lesson

6. Complete the following sentence. In a 30°-60° right triangle, the hypotenuse is always ________ the length of the side opposite the 30° angle.

7. The two legs in a 45°-45° right triangle are always congruent. Why?

Helping You Remember

8. The sides of a triangle in a 30°-60° right triangle and 45°-45° right triangle each have a special relationship. How can you remember which relationship goes with each triangle?

NAME ______________________ DATE ____________ PERIOD _____

6-4 Reading to Learn Mathematics

Classifying Quadrilaterals

Pre-Activity ***Complete the Mini Lab at the top of page 272 in your textbook. Write your answers below.***

1. Name the shape of the figures formed when you drew the diagonal. How many figures were formed?

2. You know that the sum of the angle measures of a triangle is 180°. Use this fact to find the sum of the angle measures in a quadrilateral. Explain your reasoning.

3. Find the measure of each angle of your quadrilateral. Compare the sum of these measures to the sum you found in Exercise 2.

Reading the Lesson

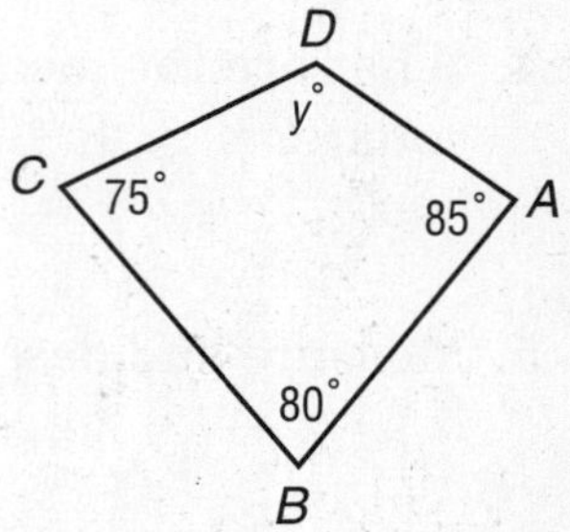

4. Explain what is done at each step in the following problem. Find the value of y in quadrilateral $ABCD$.

$m\angle A + m\angle B + m\angle C + m\angle D = 360$ ______________________

$85 + 80 + 75 + y = 360$ ______________________

$240 + y = 360$ ______________________

$\underline{-240 \qquad = -240}$ ______________________

$y = 120$ ______________________

5. How can you tell the difference between a square and a rhombus?

Helping You Remember

6. Complete the table below by placing an "x" in each box to show the correct information for each shape.

	opposite sides parallel	4 congruent sides	exactly one pair opposite sides parallel	4 right angles	opposite sides congruent
Parallelogram					
Rhombus					
Square					
Rectangle					
Trapezoid					

NAME ______________________________ DATE ____________ PERIOD _____

6-5 Reading to Learn Mathematics

Congruent Polygons

Pre-Activity ***Read the introduction at the top of page 279 in your textbook. Write your answers below.***

1. How many different triangles are shown in the Winter Stars quilt? Explain your reasoning and draw each triangle.

2. Copy the quilt and label all matching triangles with the same number, starting with 1.

Reading the Lesson

4. What does it mean for two angles to be congruent?

5. What does it mean for two polygons to be congruent?

6. What does congruence mean in your own words?

Helping You Remember

7. Make a simple drawing of the front of a house. Include at least two pairs of congruent polygons in your drawing. Then write a sentence describing the congruent polygons.

NAME ______________________ DATE ____________ PERIOD _____

6-6 Reading to Learn Mathematics

Symmetry

Pre-Activity ***Complete the Mini Lab at the top of page 286 in your textbook. Write your answers below.***

1. Draw a line down the center of your starfish outline. Then fold your paper across this line. What do you notice about the two halves?

2. Are there other lines you can draw on your outline that will produce the same result? If so, how many?

3. Place the transparency over the outline on your tracing paper. Use your pencil point at the centers of the starfish to hold the transparency in place. How many times can you rotate the transparency from its original position so that the two figures match? Do not count the original position.

4. Find the first angle of rotation by dividing 360° by the number of times the figures matched.

5. List the other angles of rotation by adding the first angle of rotation to the previous angle. Stop when you reach 360°.

Reading the Lesson

6. What is a line of symmetry?

7. What does it mean to rotate a figure?

8. After a 360° rotation about its center, a figure looks exactly as it does in its original position. Does this mean that the figure has rotational symmetry? Explain.

Helping You Remember

9. Name two everyday things that have rotational symmetry. Explain how you know these items have rotational symmetry.

NAME ______________________ DATE ____________ PERIOD _____

6-7 Reading to Learn Mathematics

Reflections

Pre-Activity ***Read the introduction at the top of page 290 in your textbook. Write your answers below.***

1. Compare the shape and size of the bird to its image in the water.

2. Compare the perpendicular distance from the water line to each of the points shown. What do you observe?

3. The points A, B, and C appear *counterclockwise* on the bird. How are these points oriented on the bird's image?

Reading the Lesson

For Exercises 4–7, read the Key Concept information on page 290 in your textbook and refer to the graph at the right that shows quadrilaterals *ABCD* and *A′B′C′D′* and a line.

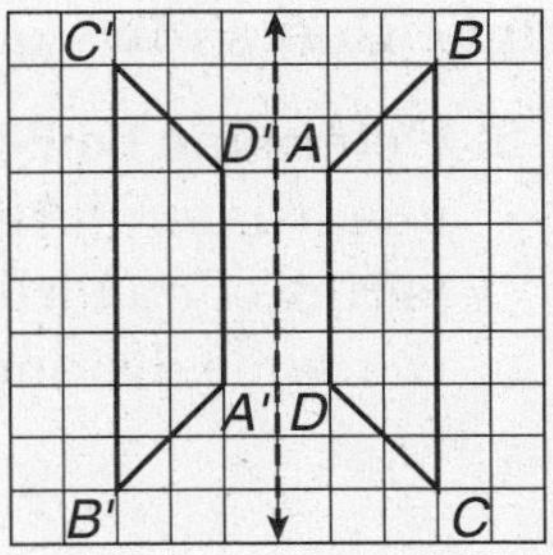

4. Find the distance of each vertex from the given line. Are corresponding vertices, such as A and A', the same distance from the line?

5. Are quadrilaterals $ABCD$ and $A'B'C'D'$ congruent? Explain.

6. Do the vertices A, B, C, and D appear clockwise or counterclockwise on quadrilateral $ABCD$? Do the vertices A', B', C', and D' appear clockwise or counterclockwise on quadrilateral $A'B'C'D'$? Do the two figures have the same orientation?

7. Is quadrilateral $A'B'C'D'$ a reflection of quadrilateral $ABCD$ over the given line? Why or why not?

Helping You Remember

8. Find two real-life examples of objects that have line symmetry. Then draw a sketch of each so that the sketch can be reflected over a line to produce a drawing of the entire object.

NAME ______________________ DATE ____________ PERIOD _____

6-8 Reading to Learn Mathematics

Translations

Pre-Activity ***Read the introduction at the top of page 296 in your textbook. Write your answers below.***

1. Describe the motion involved in moving the knight.

2. Compare the shape, size, and orientation of the knight in its original position to that of the knight in its new position.

Reading the Lesson

For Exercises 3–6, read the Key Concept information on page 296 in your textbook and refer to the graph at the right that shows quadrilaterals *ABCD* and *A´B´C´D´*.

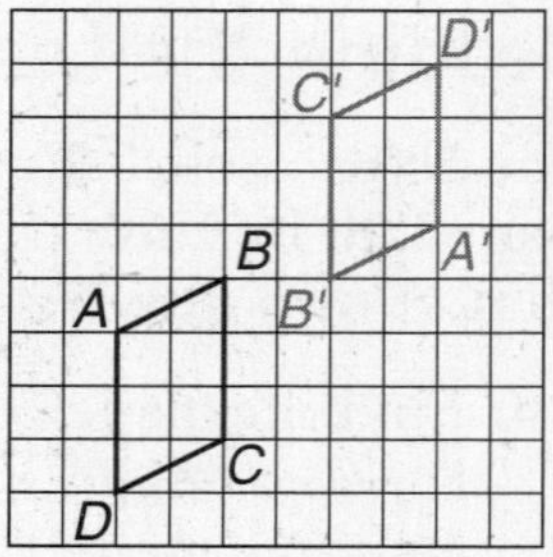

3. Determine how many units up and to the right each vertex of quadrilateral ABCD has moved to get to the corresponding vertex of quadrilateral *A´B´C´D´*. Have all the vertices moved the same distance in the same direction?

4. Are quadrilaterals *ABCD* and *A´B´C´D´* congruent? Explain.

5. Do the vertices *A*, *B*, *C*, and *D* appear clockwise or counterclockwise on quadrilateral *ABCD*? Do the vertices *A´*, *B´*, *C´*, and *D´* appear clockwise or counterclockwise on quadrilateral *A´B´C´D´*? Do the two figures have the same orientation?

6. Is quadrilateral *A´B´C´D´* a translation of quadrilateral *ABCD*? Why or why not?

Helping You Remember

7. Translations and reflections are both types of transformations. One way to remember the difference between them is to note that a translation is a slide while a reflection is a flip. Come up with your own ways to help you distinguish between the different types of transformations.

NAME ______________________________ DATE ____________ PERIOD _____

6-9 Reading to Learn Mathematics

Rotations

Pre-Activity ***Complete the Mini Lab at the top of page 300 in your textbook. Write your answers below.***

1. Measure the distances from points on the original figure and corresponding points on the image to the center of rotation. What do you observe?

2. Measure the angles formed by connecting the center of rotation to pairs of corresponding points. What do you observe?

Reading the Lesson

For Exercises 3–6, read the Key Concept information on page 300 of your textbook and refer to the graph at the right that shows quadrilaterals *ABCD* and *A′B′C′D′* and a point.

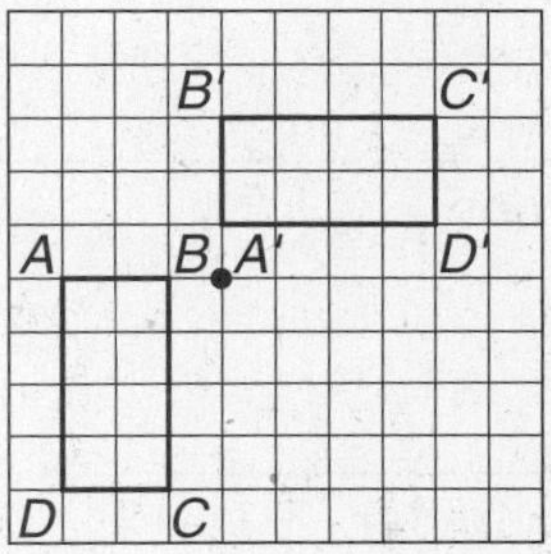

3. Find the distance of each of the vertices *A*, *B*, *A′*, and *B′* from the given point. Are corresponding vertices, such as *A* and *A′*, the same distance from the point?

4. Are quadrilaterals *ABCD* and *A′B′C′D′* congruent? Explain

5. Do the vertices *A*, *B*, *C*, and *D* appear clockwise or counterclockwise on quadrilateral *ABCD*? Do the vertices *A′*, *B′*, *C′*, and *D′* appear clockwise or counterclockwise on quadrilateral *A′B′C′D′*? Do the two figures have the same orientation?

6. Is quadrilateral *A′B′C′D′* a rotation of quadrilateral *ABCD* about the given point? Why or why not?

Helping You Remember

7. Complete the table below to help you remember the different types of transformations.

Movement	Type of Transformation
Turning about a point	
Flipping across a line	
Sliding in the plane	

CHAPTER 7

NAME ______________________________ DATE ______________ PERIOD ______

Reading to Learn Mathematics

Vocabulary Builder

This is an alphabetical list of new vocabulary terms you will learn in Chapter 7. As you study the chapter, complete each term's definition or description. Remember to add the page number where you found the term. Add this page to your math study notebook to review vocabulary at the end of the chapter.

Vocabulary Term	Found on Page	Definition/Description/Example
circumference		
complex figure		
cone		
cylinder		
diameter		
edge		
face		
lateral area		
lateral face		

NAME ______________________ DATE ____________ PERIOD _____

Reading to Learn Mathematics *(continued)*

Vocabulary Builder

Vocabulary Term	Found on Page	Definition/Description/Example
polyhedron		
precision		
prism		
pyramid		
radius		
significant digits		
slant height		
surface area		
vertex		
volume		

7-1 Reading to Learn Mathematics

Area of Parallelograms, Triangles, and Trapezoids

Pre-Activity ***Complete the Mini Lab at the top of page 314 in your textbook. Write your answers below.***

1. What dimensions are the same in each figure?

2. Compare the areas of the three figures. What do you notice?

Reading the Lesson

3. If $A = \ell w$ is the formula for the area of a rectangle, make a conjecture about the formulas for the area of a parallelogram and triangle.

4. How do you find the area of a parallelogram?

5. How can you determine the height of a parallelogram, triangle, or trapezoid?

Helping You Remember

6. Complete the table to help you remember the formulas for each figure.

Figure	Formula
parallelogram	
rectangle	
triangle	
trapezoid	

7-2 Reading to Learn Mathematics

Circumference and Area of Circles

Pre-Activity ***Complete the Mini Lab at the top of page 319 in your textbook. Write your answers below.***

1. What distance does C represent?

2. Find the ratio $\frac{C}{d}$ for this object.

3. Repeat the steps above for at least two other circular objects and compare the ratios of C to d. What do you observe?

4. Plot the data you collected as ordered pairs, (d, C). Then find the slope of a best-fit line through these points.

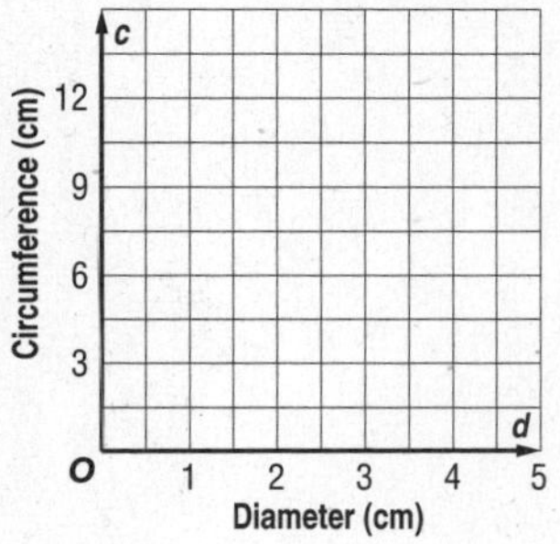

Reading the Lesson

5. Explain the difference between the radius and the diameter of a circle.

6. What is the ratio of the circumference of a circle to its diameter?

7. Explain how you find the circumference of a circle given its radius is 4 inches.

Helping You Remember

8. One way to help you remember a formula or concept is to make up a saying. For example, to remember the formula for the area of a circle you might use, "Fuzzy Wuzzy was a bear; area equals π (pi) r squared." Make up your own sayings to help you remember the formulas for the circumference and area of circles.

NAME ______________________ DATE __________ PERIOD ____

7-3 Reading to Learn Mathematics

Area of Complex Figures

Pre-Activity ***Read the introduction at the top of page 326 in your textbook. Write your answers below.***

1. Identify some of the polygons that make up the family room, nook, and foyer area shown in this floor plan.

Reading the Lesson

2. What is a complex figure?

3. What is the first step in finding the area of a complex figure?

4. Explain how to divide up the figure shown.

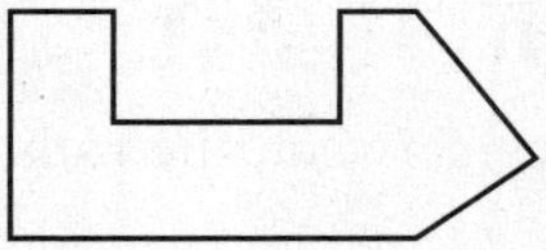

Helping You Remember

5. Look up the everyday definition of the word *complex* in a dictionary. How does the definition relate to what you learned in the lesson?

NAME ______________________________ DATE ______________ PERIOD _____

7-4 Reading to Learn Mathematics

Three-Dimensional Figures

Pre-Activity ***Read the introduction at the top of page 331 in your textbook. Write your answers below.***

1. Name the two-dimensional shapes that make up the sides of this crystal.

2. If you observed the crystal from directly above, what two-dimensional figure would you see?

3. How are two- and three-dimensional figures related?

Reading the Lesson

4. A plane is a two-dimensional flat surface that extends in all directions. What is formed when two planes intersect?

5. How can you tell the difference between a prism and pyramid?

6. Identify a three-dimensional figure that has one base that is a hexagon and six other faces that are triangles.

Helping You Remember

7. Visualize your classroom as a prism with yourself sitting in the middle of the room. What parts of the classroom represent an edge, a face, and a vertex?

7-5 Reading to Learn Mathematics

Volume of Prisms and Cylinders

Pre-Activity ***Complete the Mini Lab at the top of page 335 in your textbook. Write your answers below.***

1. Describe how the volume V of each prism is related to its length ℓ, width w, and height h.

2. Describe how the area of the base B and the height h of each prism is related to its volume V.

Reading the Lesson

3. What is another way to write the volume of a rectangular prism other than $V = Bh$?

4. What does it mean if a figure has a volume of 120 cubic centimeters?

5. Explain how finding the volume of a complex solid is similar to finding the area of a complex figure.

Helping You Remember

6. Complete the table below by filling in the correct formula.

Figure	Formula for Finding Volume
rectangular prism	
triangular prism	
cylinder	

NAME ________________________________ DATE ____________ PERIOD _____

7-6 Reading to Learn Mathematics

Volume of Pyramids and Cones

Pre-Activity ***Complete the Mini Lab at the top of page 342 in your textbook. Write your answers below.***

1. Compare the base areas and the heights of the two solids.

2. Fill the pyramid with rice, sliding a ruler across the top to level the amount. Pour the rice into the cube. Repeat until the prism is filled. How many times did you fill the pyramid in order to fill the cube?

3. What fraction of the cube's volume does one pyramid fill?

Reading the Lesson

4. How is the volume of a cone related to that of a cylinder?

5. How is the volume of a pyramid related to that of a prism?

6. Fill in the table about what you know from the diagram. Then compute the volume of the pyramid.

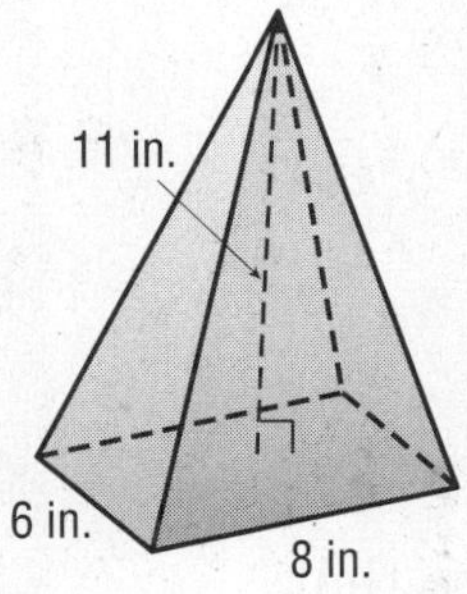

length of rectangle	
width of rectangle	
area of base	
height of pyramid	
volume of pyramid	

Helping You Remember

7. Find some examples of real-life cones. Draw a sketch of one item.

NAME ______________________ DATE ____________ PERIOD _____

7-7 Reading to Learn Mathematics

Surface Area of Prisms and Cylinders

Pre-Activity ***Complete the Mini Lab at the top of page 347 in your textbook. Write your answers below.***

1. Estimate the area in square centimeters of each face for one of your boxes. Then find the sum of these six areas.

2. Now use your ruler to measure the sides of each face. Then find the area of each face to the nearest square centimeter. Find the sum of all these areas and compare to your estimate.

3. Estimate and then find the surface areas of your other boxes.

Reading the Lesson

4. Complete the sentence with the correct numbers. When you draw a net of a triangular prism, there are ____ congruent triangular faces and ______ rectangular faces.

5. Explain how using a net helps to find the surface area of a figure.

6. If you unroll a cylinder, what does the net look like?

Helping You Remember

7. Surface area contains the word *face*. Remember to turn the object so that each side *faces* you, and no face is left out. Fill in the chart to help you remember how many faces you should be looking for in each figure.

Figure	Total Number of Faces
rectangular prism	
trianglular prism	

7-8 Reading to Learn Mathematics

Surface Area of Pyramids and Cones

Pre-Activity ***Read the introduction at the top of page 352 in your textbook. Write your answers below.***

1. How many cloth faces does this pyramid have? What shape are they?

2. How could you find the total area of the material used for the parachute?

Reading the Lesson

3. Complete the steps in finding the surface area of a pyramid.

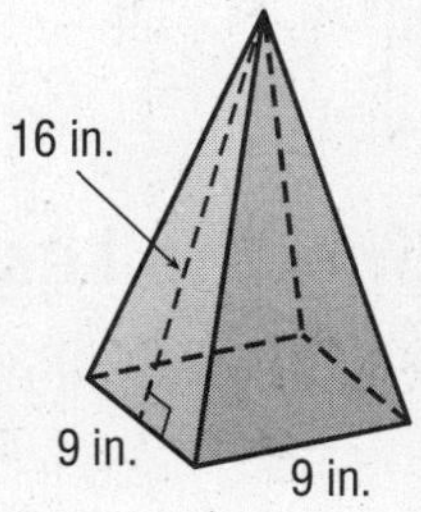

Area of each lateral face

$A = \frac{1}{2}bh$ ________________________

$A = \frac{1}{2}(9)(16)$ ________________________

$A = 72$ ________________________

There are _____ faces, so the lateral area is $4(72) =$ _____ square inches.

Area of base

$A = s^2$ ________________________

$A = 9^2$ or 81 ________________________

The surface area of the pyramid is $288 + 81 =$ _____ square inches.

4. What two areas are needed to calculate the surface area of a cone?

5. In a pyramid, what is the altitude of each face called?

Helping You Remember

6. Explain how the slant height of a cone is different from the height of the cone. Find a real-life example of a solid cone and use it to determine whether it is easier to measure the height or the slant height of a cone. Explain your reasoning.

7-9 Reading to Learn Mathematics

Precision and Significant Digits

Pre-Activity ***Read the introduction at the top of page 358 in your textbook. Write your answers below.***

1. How precisely has the daughter, Dot, measured each piece?

2. Give an example of a situation where this degree of accuracy might be appropriate.

Reading the Lesson

3. Explain precision unit.

4. How do you determine that a measurement is accurate?

5. What determines the number of significant digits?

Helping You Remember

6. Read each statement in the table. Place an X in the column in the table to make each rule true.

	Significant	*Not* Significant
All nonzero digits are...		
Zeros between two significant digits are...		
In a number with a decimal point, all zeros to the right of a nonzero digit are...		
In a number without a decimal point, any zeros to the right of the last nonzero digit are...		

NAME ______________________ DATE ____________ PERIOD _____

Reading to Learn Mathematics

Vocabulary Builder

This is an alphabetical list of new vocabulary terms you will learn in Chapter 8. As you study the chapter, complete each term's definition or description. Remember to add the page number where you found the term. Add this page to your math study notebook to review vocabulary at the end of the chapter.

Vocabulary Term	Found on Page	Definition/Description/Example
biased sample		
combination		
complementary events		
compound events		
dependent events		
experimental probability		
factorial		
Fundamental Counting Principle		
independent events		
outcome		

NAME ______________________ DATE ____________ PERIOD _____

Reading to Learn Mathematics

Vocabulary Builder *(continued)*

Vocabulary Term	Found on Page	Definition/Description/Example
permutation		
population		
probability		
random		
sample		
stratified random sample		
systematic random sample		
theoretical probability		
tree diagram		
unbiased sample		

NAME ______________________________ DATE ____________ PERIOD _____

8-1 Reading to Learn Mathematics

Probability of Simple Events

Pre-Activity ***Read the introduction at the top of page 374 in your textbook. Write your answers below.***

1. Write the ratio that compares the number of double tiles to the total number of tiles.

2. What percent of the tiles are doubles?

3. Write a fraction in simplest form that represents the part of the tiles that are doubles.

4. Write a decimal that represents the part of the tiles that are doubles.

5. Suppose you pick a domino without looking at the spots. Would you be more likely to pick a tile that is a double or one that is not a double? Explain.

Reading the Lesson

6. How is *probability* defined on page 374 of your textbook?

7. If a spinner is divided into 8 equal parts, and one of the first eight letters of the alphabet are placed in each section, what is the sample space?

8. If an event has a probability of 0.1, what does this mean?

9. If you are given the probability of an event happening, how can you find the probability of the event *not* happening?

Helping You Remember

10. Work with a partner to design a spinner that has a probability of spinning red as $\frac{2}{5}$, blue as $\frac{4}{10}$, and green as $\frac{3}{15}$.

NAME ______________________ DATE __________ PERIOD _____

8-2 Reading to Learn Mathematics

Counting Outcomes

Pre-Activity ***Read the introduction at the top of page 380 in your textbook. Write your answers below.***

1. How many different styles are available?
2. How many different colors are available?
3. How many different sizes are available?
4. Make an organized list to determine how many different bicycles are available.

Reading the Lesson

5. Complete the tree diagram shown below for how many boys and how many girls are likely to be in a family of three children.

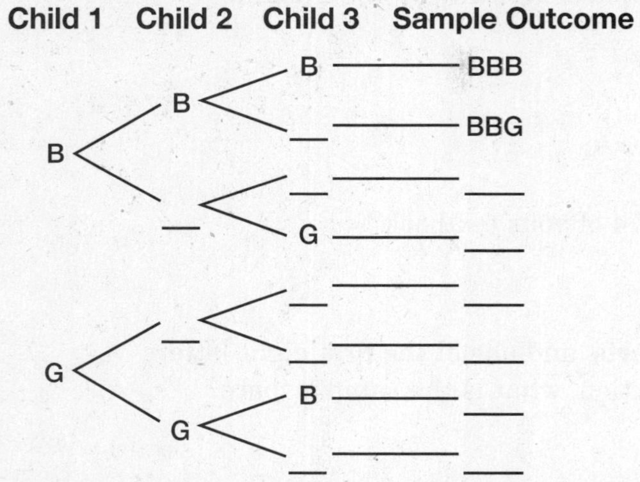

6. Explain when you would not want to use a tree diagram.

Helping You Remember

7. Make a simple lunch menu. Include at least two sandwich choices, 3 side choices, and 3 dessert choices. Then write a sentence describing how you could determine the number of different meals that can be made by choosing only one item from each category.

8-3 Reading to Learn Mathematics

Permutations

Pre-Activity *Complete the Mini Lab at the top of page 384 in your textbook. Write your answers below.*

1. How many different arrangements did you make?

2. How many different game pieces could you pick for the first place?

3. Once you pick the first-place game piece, how many game pieces could you pick for the second place?

4. Use the Fundamental Counting Principle to determine the number of arrangements for first and second places.

5. How do the numbers in Exercises 1 and 4 compare?

Reading the Lesson

6. Complete the sentence: An arrangement or listing in which order is important is called a ________________.

7. What does the notation $P(14, 4)$ represent?

8. What is another way to write $5 \cdot 4 \cdot 3 \cdot 2 \cdot 1$?

9. What does 6! mean?

10. A security system has a number pad with 9 digits. How many three-number passwords are available if a digit cannot be repeated? If a digit can be repeated, how many passwords are available?

Helping You Remember

11. An arrangement or listing in which order is important is called a permutation. Think about the word permutation. How can you remember that order is important? Give an example of when order would be important.

NAME ______________________________ DATE ____________ PERIOD _____

8-4

Reading to Learn Mathematics

Combinations

Pre-Activity ***Complete the Mini Lab at the top of page 388 in your textbook. Write your answers below.***

1. How many different handshakes did you record?

2. Find $P(6, 2)$.

3. Is the number of handshakes equal to $P(6, 2)$? Explain.

Reading the Lesson

4. What is the difference between a permutation and a combination?

5. Fill in the blanks. Find $C(9, 4)$.

$$C(9, 4) = \frac{P(9, 4)}{\_\_\_!}$$

$$= \frac{9 \cdot \_\_\_ \cdot \_\_\_ \cdot \_\_\_}{4 \cdot \_\_\_ \cdot \_\_\_ \cdot \_\_\_}$$

$$= \_\_\_\_$$

6. Are there more combinations or permutations of 3 people chosen from a group of 6 people? Explain.

Helping You Remember

7. You and a friend walk into a fast-food restaurant and each order the same combination meal. You receive your meal in the order of hamburger, french fries, and drink. Your friend receives her meal in the order of french fries, drink, and hamburger. You each ordered the same combination meal but the order in which you received your food did not matter. Give another example for which order does not matter.

NAME ______________________ DATE ____________ PERIOD ______

8-5 Reading to Learn Mathematics

Probability of Compound Events

Pre-Activity ***Read the introduction at the top of page 396 in your textbook. Write your answers below.***

1. A player rolls the number cube. What is $P(\text{odd})$?

2. The player spins the spinner. What is $P(\text{red})$?

3. What is the product of the probabilities in Exercises 1 and 2?

4. On a separate piece of paper, draw a tree diagram to determine the probability that the player will get an odd number and red.

5. Compare your answers for Exercises 3 and 4.

Reading the Lesson

6. What is a compound event?

7. Are the events of spinning a spinner and rolling a number cube independent events? Why or why not?

8. Explain how to find the probability of two independent events.

Helping You Remember

9. Look up the everyday definitions of the words *dependent* and *independent* in a dictionary. How does each definition relate to what you have learned in this lesson?

NAME ______________________ DATE ____________ PERIOD _____

8-6 Reading to Learn Mathematics

Experimental Probability

Pre-Activity ***Complete the Mini Lab at the top of page 400 in your textbook. Write your answers below.***

1. Compute the ratio $\frac{\text{number of times color was drawn}}{\text{total number of draws}}$ for each color of marble.

2. Is it possible to have a certain color marble in the bag and never draw that color?

3. Open the bag and count the marbles. Find the ratio $\frac{\text{number of each color marble}}{\text{total number of marbles}}$ for each color of marble.

4. Are the ratios in Exercises 1 and 3 the same? Explain why or why not.

Reading the Lesson

5. The table at the right shows the results of a survey.

How many people bought balloons? __________

How many people were surveyed? __________

What is the experimental probability that a person surveyed preferred balloons?

Item	Number of People
balloons	75
cards	15
decorations	25
cake	50

6. A bag contains 15 red marbles, 25 purple marbles, and 10 yellow marbles. Describe an experiment that you could conduct with the marbles to find an experimental probability.

Helping You Remember

7. One way to remember the difference between experimental probability and theoretical probability is to note that experimental probability is based on an experiment and theoretical probability is based on what should happen in theory. Look in a newspaper and find an example of each type of probability.

NAME ______________________ DATE ____________ PERIOD _____

8-7 Reading to Learn Mathematics

Using Sampling to Predict

Pre-Activity ***Read the introduction at the top of page 406 in your textbook. Write your answers below.***

1. Suppose she decides to survey a group of people at a rock concert. Do you think the results would represent all of the people in the listening area? Explain.

2. Suppose she decides to survey students at your middle school. Do you think the results would represent all of the people in the listening area? Explain.

3. Suppose she decides to call every 100th household in the telephone book. Do you think the results would represent all of the people in the listening area? Explain.

Reading the Lesson

4. What is the first step in conducting a survey?

5. When you conduct a survey by asking ten students selected at random from each grade at your school what their favorite class is, what type of random sample have you taken?

6. A grocery store owner asks the shoppers in his store where they prefer to shop for groceries. What type of sample has he conducted?

Helping You Remember

7. Working with a partner, complete the table below with an example of each type of sample.

Type of Sample	Example
systematic random	
simple random	
stratified random	
convenience	
voluntary response	

CHAPTER 9

NAME ______________________ DATE ____________ PERIOD _____

Reading to Learn Mathematics

Vocabulary Builder

This is an alphabetical list of new vocabulary terms you will learn in Chapter 9. As you study the chapter, complete each term's definition or description. Remember to add the page number where you found the term. Add these pages to your math study notebook to review vocabulary at the end of the chapter.

Vocabulary Term	Found on Page	Definition/Description/Example
box-and-whisker plot		
circle graph		
column		
dimensions		
element		
histogram		
interquartile range		
lower quartile		
matrix [MAE-triks]		

NAME ______________________________ DATE ______________ PERIOD _____

Reading to Learn Mathematics *(continued)*

Vocabulary Builder

Vocabulary Term	Found on Page	Definition/Description/Example
mean		
measures of central tendency		
measures of variation		
median		
mode		
outlier		
quartiles		
range		
row		
upper quartile		

NAME ______________________ DATE ____________ PERIOD ______

9-1 Reading to Learn Mathematics

Histograms

Pre-Activity ***Read the introduction at the top of page 420 in your textbook. Write your answers below.***

1. What do you notice about the price intervals?

2. What does each tally mark represent?

3. How is the frequency for each price range determined?

Reading the Lesson

4. Explain the difference between a bar graph and a histogram.

5. What does a histogram display?

6. Can the data in the table at the right be used to draw a histogram? Explain.

Eye Color in an Eighth-Grade Class		
Color	**Tally**	**Frequency**
Blue	~~||||~~ ~~||||~~ ||	12
Brown	~~||||~~ ||||	9
Green	|||	3

7. What is wrong with the histogram at the right?

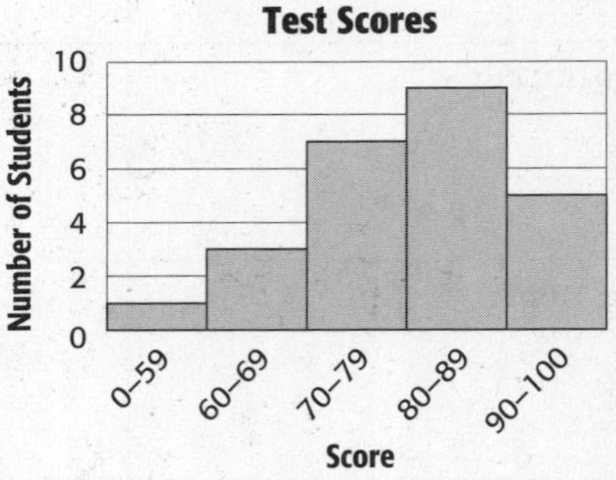

Helping You Remember

8. Work with a partner. Have one partner create a frequency table. Have the other partner draw a histogram from the table.

9-2 Reading to Learn Mathematics

Circle Graphs

Pre-Activity ***Read the introduction at the top of page 426 in your textbook. Write your answers below.***

1. What percent of the public roads are owned by the counties?

2. What government owns 19.6% of the public roads?

3. How do you know that all types of government have been accounted for?

Reading the Lesson

4. Complete the sentence: A circle graph compares ____________ to the ____________.

5. When drawing a circle graph from given percents, explain how the degrees for each section are determined.

6. What must you do before you can construct a circle graph for the data in the table at the right?

Favorite Shoes	
Dress	17
Sandals	21
Sneakers	42

Helping You Remember

7. Can the data in the table at the right be used to make a circle graph? Explain. What must be true of the percents used to make a circle graph?

Color of T-Shirt Worn	
Blue	32%
Red	19%
Gray	6%

NAME ______________________________ DATE ____________ PERIOD _____

9-3 Reading to Learn Mathematics

Choosing an Appropriate Display

Pre-Activity ***Read the introduction at the top of page 430 in your textbook. Write your answers below.***

1. Which display(s) show all of the individual test scores?

2. Do any of the displays allow you to find the test score of a certain student? If not, what type of display would show this type of information?

Reading the Lesson

3. Name three different ways to display data.

4. What two questions should you ask yourself when determining which type of display to use?

Helping You Remember

Choose the letter that best matches the type of display to its use.

5. Circle Graph	**a.** shows the frequency of data that has been organized into equal intervals
6. Line Graph	**b.** lists all individual numerical data in a condensed form
7. Pictograph	**c.** shows the number of items in specific categories in the data using bars
8. Table	**d.** compares part to a whole
9. Bar Graph	**e.** may list all the data individually or by groups
10. Histogram	**f.** shows change over a period of time
11. Line Plot	**g.** shows the number of items in specific categories using symbols to represent a quantity
12. Stem-and-Leaf Plot	**h.** shows how many times each number occurs in the data

NAME ______________________ DATE ____________ PERIOD _____

9-4 Reading to Learn Mathematics

Measures of Central Tendency

Pre-Activity ***Read the introduction at the top of page 435 in your textbook. Write your answers below.***

1. What is the *average* number of days for these nine countries?

2. Order the numbers from least to greatest. What is the middle number in your list?

3. What number(s) appear more than once?

4. Which of the number or numbers from Exercises 1–3 might be representative of the set of data? Explain.

Reading the Lesson

5. Name the three most common measures of central tendency.

6. Explain in your own words how to find the mean of a data set.

7. When finding the median, what first must be done to the set of data?

Helping You Remember

8. Think about the hours of television you have watched each day in the past week. List the times, and find their mean, median, and mode. Which measure of central tendency best represents the data? Compare your findings with your classmates.

9-5 Reading to Learn Mathematics

Measures of Variation

Pre-Activity ***Read the introduction at the top of page 442 in your textbook. Write your answers below.***

1. What is the greatest number of hours spent online?

2. What is the least number of hours spent online?

3. Find the difference between the greatest number and the least number of hours spent online.

4. Write a sentence explaining what the answer to Exercise 3 says about the data.

Reading the Lesson

5. When given a set of data, explain how to find the range.

6. What is the interquartile range?

7. Describe the steps for finding the limits for outliers of a set of data.

Helping You Remember

8. Explain how the word "quarter" and "quartile" are similar. Explain how they are different.

NAME ______________________ DATE ____________ PERIOD _____

9-6 Reading to Learn Mathematics

Box-and-Whisker Plots

Pre-Activity ***Read the introduction at the top of page 446 in your textbook. Write your answers below.***

1. What is the least value in the data?

2. What is the lower quartile of the data?

3. What is the median of the data?

4. What is the upper quartile of the data?

5. What is the greatest value in the data?

6. Name any outliers.

Reading the Lesson

7. Explain the parts of the box-and-whisker plot in your own words.

8. Complete the sentence: A box-and-whisker plot separates a set of data into ______ parts, each containing ________ of the data.

9. What does the length of a whisker tell you about the data?

Helping You Remember

10. What five key numbers are necessary to draw a box-and-whisker plot? Identify the point corresponding to each of these numbers on the graph below.

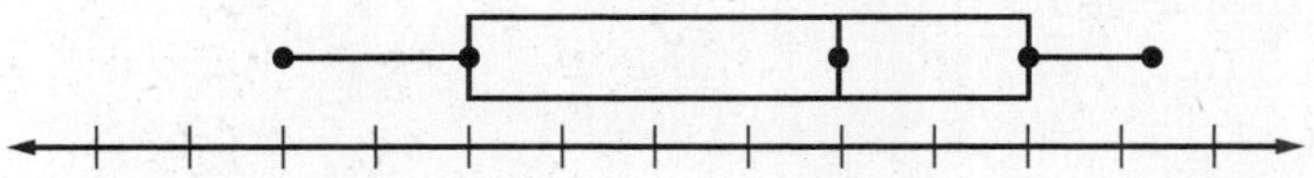

9-7 Reading to Learn Mathematics

Misleading Graphs and Statistics

Pre-Activity ***Read the introduction at the top of page 450 in your textbook. Write your answers below.***

1. Do both graphs show the same data?

2. Which graph seems to show the greater increase in spending? Why?

Reading the Lesson

3. Give an example of how a graph could be misleading.

4. When writing an employment ad for an automotive dealership, would it be best to use the mean, median, or mode of the number of cars sold to encourage a commissioned salesperson to apply for the job?

Wagner Automotive Sales			
Month	**Number of Cars Sold**	**Month**	**Number of Cars Sold**
January	16	July	44
February	5	August	40
March	34	September	38
April	49	October	45
May	47	November	48
June	79	December	38

Helping You Remember

5. Work with a partner. Have your partner draw a set of graphs that are misleading. Tell your partner what makes the graphs misleading and redraw them to accurately display the data.

NAME ______________________ DATE ____________ PERIOD _____

9-8 Reading to Learn Mathematics

Matrices

Pre-Activity ***Read the introduction at the top of page 454 in your textbook. Write your answers below.***

1. How many continents are listed?

2. How many temperatures are given for each continent?

3. What does the number −129 represent?

4. What does the number 136 represent?

Reading the Lesson

5. Describe the matrix in terms of its dimensions.

$$\begin{bmatrix} 26 & 5 \\ 45 & 4 \\ 36 & 9 \\ 14 & 3 \end{bmatrix}$$

6. Before you can add or subtract two matrices, what must be the same? Explain.

7. Explain the difference between the words *row* and *column*.

8. What is each number of a matrix called?

Helping You Remember

9. Take two 2 by 2 matrices and place blue circles around the numbers in row 1 column 1, place red circles around the numbers in row 1 column 2, green circles around the numbers in row 2 column 1, and purple circles around the numbers in row 2 column 2. To add or subtract the matrices, add or subtract the numbers circled in the same color.

CHAPTER 10

NAME ______________________ DATE ____________ PERIOD _____

Reading to Learn Mathematics

Vocabulary Builder

This is an alphabetical list of new vocabulary terms you will learn in Chapter 10. As you study the chapter, complete each term's definition or description. Remember to add the page number where you found the term. Add this page to your math study notebook to review vocabulary at the end of the chapter.

Vocabulary Term	Found on Page	Definition/Description/Example
coefficient		
constant		
equivalent expressions		
like terms		
simplest form		
simplifying the expression		
term		
two-step equation		

NAME ______________________ DATE ____________ PERIOD ______

10-1 Reading to Learn Mathematics

Simplifying Algebraic Expressions

Pre-Activity ***Do the Mini Lab at the top of page 469 in your textbook. Write your answers below.***

1. Choose two positive and one negative value for x. Then evaluate $2(x + 3)$ and $2x + 6$ for each of these values. What do you notice?

2. Use algebra tiles to rewrite the expression $3(x - 2)$. (*Hint:* Use one green x-tile and 2 red 1-tiles to represent $x - 2$.)

Reading the Lesson

3. When is the Distributive Property used to simplify an algebraic expression?

4. Explain how to simplify the expression $5(x - 3)$.

5. Explain what it means for two expressions to be equivalent.

6. Give an example of an expression containing three terms, one of which is a constant.

Helping You Remember

7. One of your classmates was absent from school today and has not studied the lesson. Write a letter to your classmate explaining how to simplify an expression and how to identify terms and constants.

NAME ______________________ DATE ____________ PERIOD _____

10-2 Reading to Learn Mathematics

Solving Two-Step Equations

Pre-Activity ***Read the introduction at the top of page 474 in your textbook. Write your answers below.***

1. Explain how you could use the work backward strategy to find the cost of each hardback book. Then find the cost.

Reading the Lesson

2. Define *two-step equation.*

Determine whether each equation is a two-step equation. Explain.

3. $n + 8 = 21$

4. $2x + 1 = 15$

5. $\frac{c}{4} = 6.5$

What is the first step in solving each equation?

6. $3y - 2 = 16$

7. $5 - 6x = -19$

8. $-2p + 11 = 7$

Helping You Remember

9. Draw a diagram that shows how the equation $2x + 3 = 8$ can be modeled using algebra tiles.

NAME ______________________ DATE ____________ PERIOD _____

10-3 Reading to Learn Mathematics

Writing Two-Step Equations

Pre-Activity ***Read the introduction at the top of page 478 in your textbook. Write your answers below.***

1. Let n represent the number of payments. Write an expression that represents the amount of the loan paid after n payments.

2. Write and solve an equation to find the number of payments you will have to make in order to pay off your loan.

3. What type of equation did you write for Exercise 2? Explain your reasoning.

Reading the Lesson

Jennifer bought 3 CDs, each having the same price. Her total for the purchase was $51.84, which includes $3.84 in sales tax. Find the price of each CD.

4. Explain how to define the variable in the problem. Then define the variable.

5. The next step is to write an equation for the problem. Assuming that the total, 51.84, will be on the right side of the equals sign by itself, determine which two operations will be represented on the left side of the equals sign. Which is performed first? Explain.

6. Complete the equation. Then solve it. How much does each CD cost?
________ = 51.84

Helping You Remember

7. Work with a partner. Have one partner write a word problem that involves a two-step equation and solve it. Have the other partner check the solution. Then have partners switch tasks.

NAME ______________________ DATE ____________ PERIOD _____

10-4 Reading to Learn Mathematics

Solving Equations with Variables on Each Side

Pre-Activity ***Read the introduction at the top of page 484 in your textbook. Write your answers below.***

1. Use the table. Continue filling in rows to find how long it will take you to catch up to your friend. If necessary, continue the table on another piece of paper

Time (s)	Friend's Distance (m)	Your Distance (m)
0	$15 + 5(0) = 15$	$6(0) = 0$
1	$15 + 5(1) = 20$	$6(1) = 6$
2	$15 + 5(2) = 25$	$6(2) = 12$
3	$15 + 5(3) = 30$	$6(3) = 18$

2. Write an expression for your distance after x seconds.

3. Write an expression for your friend's distance after x seconds.

4. What is true about the distances you and your friend have gone when you catch up to your friend?

5. Write an equation that could be used to find how long it will take for you to catch up to your friend.

Reading the Lesson

6. What is the first step in solving an equation with variables on each side?

7. What does it mean to *isolate* the variable when solving an equation?

Determine whether the variable is isolated in each equation. Then determine whether the equation is solved for the variable.

8. $c = 8$

9. $3x + 1 = 7$

10. $5d = 15$

Helping You Remember

11. Create a general set of guidelines to solve any type of equation.

NAME ______________________________ DATE ____________ PERIOD _____

10-5

Reading to Learn Mathematics

Inequalities

Pre-Activity ***Read the introduction at the top of page 492 in your textbook. Write your answers below.***

1. Name three truck heights that can safely pass on a road where the first sign is posted. Can a truck that is 10 feet 6 inches tall pass? Explain.

2. Name three speeds that are legal according to the second sign. Is a car traveling at 45 miles per hour driving at a legal speed? Explain.

Reading the Lesson

3. Complete the table by providing the symbol used to represent each phrase.

Phrase	Symbol	Phrase	Symbol
is greater than		is at most	
is at least		exceeds	
is fewer than		is less than or equal to	
is more than		is no less than	

4. Explain the difference between an open and a closed circle on the graph of an inequality.

5. What does the arrow to the right or to the left indicate on the graph of an inequality?

6. Describe how to graph $x > 7$.

7. Describe how to graph $x \leq -6$.

Helping You Remember

8. Use a newspaper to find real-world situations in which relationships between quantities are described by phrases like *no more than, at least, greater than*, and *at most*.

NAME ______________________ DATE ____________ PERIOD ______

10-6 Reading to Learn Mathematics

Solving Inequalities by Adding or Subtracting

Pre-Activity ***Read the introduction at the top of page 496 in your textbook. Write your answers below.***

1. Add 10 to each side of the inequality $13 < 16$. Write the resulting inequality and decide whether it is *true* or *false*.

2. Was Victoria's dad younger or older than Victoria's mom 13 years ago? Explain your reasoning using an inequality.

Reading the Lesson

3. How are solving an inequality and solving an equation similar?

4. Explain what solving an equality means.

5. Are $x \leq 7$ and $7 \leq x$ equivalent inequalities? Explain.

6. Are $x > -2$ and $-2 < x$ equivalent inequalities? Explain.

7. Write an inequality equivalent to $n < 5$, but use the $>$ symbol.

Helping You Remember

8. Look up *inequality* in a dictionary. How does its meaning relate to what you have learned in this lesson?

Reading to Learn Mathematics

Solving Inequalities by Multiplying or Dividing

Pre-Activity ***Read the introduction at the top of page 500 in your textbook. Write your answers below.***

1. Divide each side of the inequality $80 < 150$ by 2. Write the resulting inequality and decide whether it is *true* or *false*.

2. Would the cost of three pairs of basketball shoes be greater or less than the cost of three pairs of running shoes all sold at the regular price? Explain your reasoning using an inequality.

Reading the Lesson

What is the first step in solving each inequality? How is the inequality symbol affected by this first step? Explain.

3. $\frac{x}{3} > 2$

4. $-5x \leq 60$

5. $3x + 4 \geq 22$

6. Describe the conditions under which you reverse the direction of the inequality symbol when solving an inequality.

Helping You Remember

7. You can often forget to change the direction of the inequality symbol when multiplying or dividing by a negative number. The *proposed* solution to the inequality $-9x > -27$ is $x < 3$. Check the values at and near the boundary of the solution by verifying that
   - 3 is the solution of the related equation
   - 2 is a solution to the inequality
   - 4 is not a solution of the inequality

   Now create your own inequality and give three different points to check. A point that is a solution of the related equation, a point that is a solution to the inequality, and one point that is not the solution of the inequality.

CHAPTER 11

NAME ______________________________ DATE ______________ PERIOD ______

Reading to Learn Mathematics

Vocabulary Builder

This is an alphabetical list of new vocabulary terms you will learn in Chapter 11. As you study the chapter, complete each term's definition or description. Remember to add the page number where you found the term. Add this page to your math study notebook to review vocabulary at the end of the chapter.

Vocabulary Term	Found on Page	Definition/Description/Example
arithmetic sequence [air-ith-MEH-tik]		
best-fit line		
boundary		
common difference		
common ratio		
dependent variable		
domain		
function		
function table		
geometric sequence [je-o-MET-rik]		
half plane		

NAME ______________________ DATE ____________ PERIOD _____

Reading to Learn Mathematics

Vocabulary Builder *(continued)*

Vocabulary Term	Found on Page	Definition/Description/Example
independent variable		
linear function		
range		
scatter plot		
sequence		
slope formula		
slope-intercept form		
substitution		
system of equations		
term		
x-intercept		
y-intercept		

NAME ______________________________ DATE ____________ PERIOD _____

11-1 Reading to Learn Mathematics

Sequences

Pre-Activity ***Complete the Mini Lab at the top of page 512 in your textbook. Write your answers below.***

1. Continue the pattern for 4, 5, and 6 triangles. How many toothpicks are needed for each case?

2. Study the pattern of numbers. How many toothpicks will you need for 7 triangles?

3. Continue the pattern for 4, 5, and 6 squares. How many toothpicks are needed for each case?

4. How many toothpicks will you need for 7 squares?

Reading the Lesson

5. Explain how to determine whether a sequence is arithmetic.

6. Determine whether the sequence 4, 6, 9, 13, 18, . . . is arithmetic. Explain.

7. Explain how to determine whether a sequence is geometric.

8. Determine whether the sequence 2, 6, 18, 54, 162, . . . is geometric. Explain.

Helping You Remember

9. Work with a partner. Have one partner write three sequences of numbers, one that is arithmetic, one that is geometric, and one that is neither arithmetic nor geometric. Have the other partner identify the type of sequence and the common difference or ratio.

NAME ______________________________ DATE ______________ PERIOD _____

11-2 Reading to Learn Mathematics

Functions

Pre-Activity ***Read the introduction at the top of page 517 in your textbook. Write your answers below.***

1. Complete the table at the right.

Dog's Age	Equivalent Human Age
1	7
2	14
3	
4	
5	

2. If a dog is 6 years old, what is its equivalent human age?

3. Explain how to find the equivalent human age of a dog that is 10 years old.

Reading the Lesson

4. If $f(x) = x + 5$, explain how to find $f(2)$. Then find $f(2)$.

5. Identify the input value and the output value in Exercise 4.

6. Define *domain*. What number in Exercise 4 is part of the domain?

7. Explain why the output value is called the dependent variable. What represents the dependent variable in the function $f(x) = x + 5$?

Helping You Remember

8. When looking at the word *domain*, you see the word "in" located at the end of the word. This is a way to remember that the domain is the set of *in*put values. Find a way to remember that the range is the set of output values.

NAME ______________________ DATE ____________ PERIOD _____

11-3 Reading to Learn Mathematics

Graphing Linear Functions

Pre-Activity ***Read the introduction at the top of page 522 in your textbook. Write your answers below.***

1. Complete the following function table.

Input x	Rule $1.5x$	Output y	(Input, Output) (x, y)
1	1.5(1)	1.5	(1, 1.5)
2	1.5(2)		
3			
4			

2. Graph the ordered pairs on a coordinate plane.

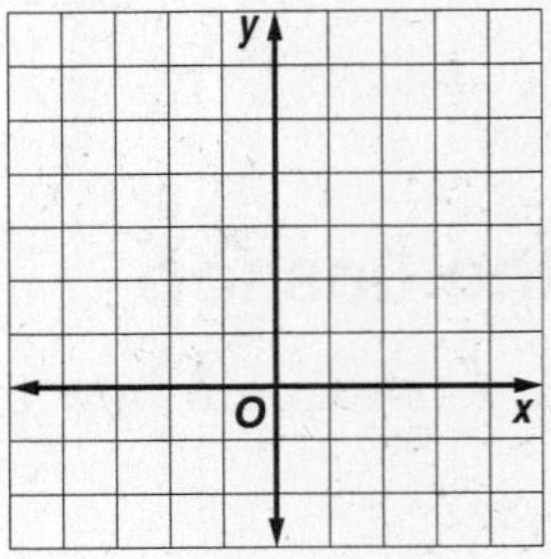

3. What do you notice about the points on your graph?

Reading the Lesson

4. In your own words, explain how to graph a function.

5. Explain how to find the x-intercept of the graph of a linear function. Then find the x-intercept of $y = 2x + 8$.

6. Explain how to find the y-intercept of the graph of a linear function. Then find the y-intercept of $y = 2x + 8$.

Helping You Remember

7. Think of a gas pump with prices for regular and super gasoline. When the same amount of gas is being pumped into a tank, how does the price per gallon affect the total cost of the gas?

NAME ______________________________ DATE ____________ PERIOD ______

11-4 Reading to Learn Mathematics

The Slope Formula

Pre-Activity ***Do the Mini Lab at the top of page 526 in your textbook. Write your answers below.***

1. Find the slope of the line by counting units of vertical and horizontal change.

2. Subtract the *y*-coordinate of *A* from the *y*-coordinate of *B*. Call this value *t*.

3. Subtract the *x*-coordinate of *A* from the *x*-coordinate of *B*. Call this value *s*.

4. Write the ratio $\frac{t}{s}$. Compare the slope of the line with $\frac{t}{s}$.

Reading the Lesson

5. A line passes through the points $A(-1, -5)$, $B(0, -1)$, $C(1, 3)$, and $D(2, 7)$. Does it matter which two points you use to find the slope using the slope formula? Explain.

6. Suppose you choose to find the slope of the line in Exercise 5 using points $C(1, 3)$ and $D(2, 7)$. If your numerator after substitution into the slope formula is $3 - 7$, what should be your denominator? Explain.

7. Explain the difference between $\frac{0}{3}$ and $\frac{3}{0}$.

Helping You Remember

8. Fill in the table with the appropriate term, *positive*, *negative*, *zero*, or *undefined*.

Description of Line	Slope
When going from left to right, the line slants upward.	
The line is horizontal.	
The line is vertical.	
When going left to right, the line slants downward.	

NAME ______________________________ DATE ____________ PERIOD _____

11-5 Reading to Learn Mathematics

Slope-Intercept Form

Pre-Activity ***Complete the Mini Lab at the top of page 533 in your textbook. Write your answers below.***

1. Use the graphs to find the slope and y-intercept of each line. Complete the table.

Equation	Slope	y-intercept
$y = 3x + 2$		
$y = \frac{1}{4}x + (-1)$		
$y = -2x + 3$		

2. Compare each equation with the value of its slope. What do you notice?

3. Compare each equation with its y-intercept. What do you notice?

Reading the Lesson

4. In the formula $y = mx + b$, what do the letters m and b represent?

Identify the slope and the y-intercept of the graph of each equation.

5. $y = -3x + 4$

6. $y = \frac{2}{3}x - 7$

7. How can you find the slope and the y-intercept of the graph of $x + y = 8$?

8. If you know the y-intercept of a line is 4 and that the slope is $-\frac{3}{2}$, how do you graph the line?

Helping You Remember

9. Work with a partner. Using a coordinate grid, take turns graphing lines and identifying the slope and y-intercept of each graph.

NAME ______________________________ DATE ____________ PERIOD _____

11-6 Reading to Learn Mathematics

Scatter Plots

Pre-Activity ***Complete the Mini Lab at the top of page 539 in your textbook. Write your answers below.***

1. Graph each of the ordered pairs listed on the chalkboard.

2. Examine the graph. Do you think there is a relationship between height and arm span? Explain.

Reading the Lesson

3. How is a scatter plot different from the graph of a linear function?

4. What pattern would you expect to see in a scatter plot that shows a positive relationship?

5. What pattern would you expect to see in a scatter plot that shows a negative relationship?

6. Would you expect a scatter plot to show a *positive*, *negative*, or *no relationship* between the population of a state and its number of representatives in the U.S. Congress? Explain.

Helping You Remember

7. Using a newspaper or magazine, find an article with data given. Plot the data on a coordinate plane and identify whether the data has a positive, negative, or no relationship.

11-7 Reading to Learn Mathematics

Graphing Systems of Equations

Pre-Activity ***Read the introduction at the top of page 544 in your textbook. Write your answers below.***

1. Graph both of the equations on a coordinate plane.

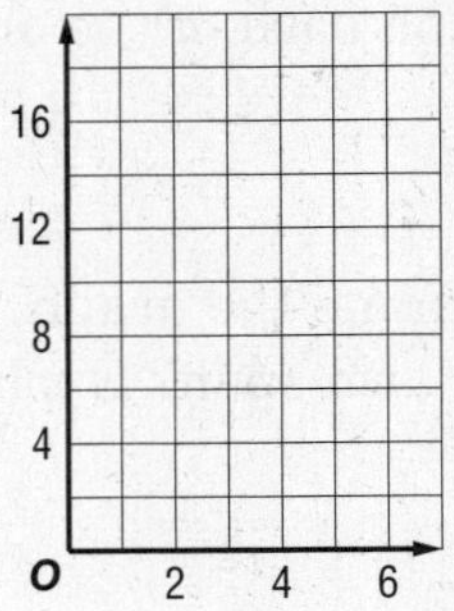

2. What are the coordinates of the point where the two lines intersect? What does this point represent?

Reading the Lesson

3. Explain how to find the solution of a system of two equations by graphing.

4. Under what conditions will a system of two linear equations have infinitely many solutions?

5. Under what conditions will a system of two linear equations have no solution?

6. What is the first step in solving the system $y = 6x + 2$ and $x = -1$ by substitution?

7. How do you check a solution of a system of two equations?

Helping You Remember

8. Use a clean sheet of paper and Examples 3 and 4 on pages 545 in your textbook. Starting with Example 3, cover everything up in the example with your paper except the title and its question. Now try to work the problem without looking at the book. Then compare your work to the work in the book. Repeat this with Example 4.

11-8 Reading to Learn Mathematics

Graphing Linear Inequalities

Pre-Activity ***Read the introduction at the top of page 548 in your textbook. Write your answers below.***

1. Use the graph to list three different combinations of CDs and books that Sabrina can purchase for $16.

2. Suppose Sabrina wants to spend less than $16. Substitute (2, 7), (4, 2), (5, 12), and (7, 5) in $2x + y < 16$. Which values make the inequality true?

3. Which region do you think represents $2x + y < 16$?

4. Suppose Sabrina can spend more than $16. Substitute (2, 7), (4, 2), (5, 12), and (7, 5) in $2x + y > 16$. Which values make the inequality true?

5. Which region do you think represents $2x + y > 16$?

Reading the Lesson

6. Why is graphing a line an important skill to have in order to graph a linear inequality?

7. What does it mean when you have a dashed boundary line? a solid boundary line?

8. Explain how you know which half plane should be shaded.

Helping You Remember

9. Complete the table to help you remember the types of lines and where to shade when graphing inequalities.

Inequality Symbol	Dashed or Solid Line
$<$	
$\geq$	
$\leq$	
$>$	

NAME ______________________________ DATE ____________ PERIOD _____

CHAPTER 12

Reading to Learn Mathematics

Vocabulary Builder

This is an alphabetical list of new vocabulary terms you will learn in Chapter 12. As you study the chapter, complete each term's definition or description. Remember to add the page number where you found the term. Add these pages to your math study notebook to review vocabulary at the end of the chapter.

Vocabulary Term	Found on Page	Definition/Description/Example
monomial		
nonlinear function		
polynomial		
quadratic function		

NAME ______________________________ DATE ______________ PERIOD _____

Reading to Learn Mathematics

Linear and Nonlinear Functions

Pre-Activity *Read the introduction at the top of page 560 in your textbook. Write your answers below.*

1. During its ascent, did the rocket travel the same distance each second? Explain.

2. During its descent, did the rocket travel the same distance each second? Explain.

3. Graph the data whose ordered pairs are (time, height) for the rocket's ascent and descent on separate axes. Connect the points with a straight line or smooth curve. Then compare the two graphs.

Reading the Lesson

4. How does the rate of change of a nonlinear function differ from the rate of change of a linear function?

5. Determine whether the table represents a linear or nonlinear function. Explain.

x	3	7	11	15
y	46	35	24	13

6. How can you distinguish the equations of linear functions from the equations of nonlinear functions?

Helping You Remember

7. Using a newspaper or magazine, find one example of a linear graph and one example of a nonlinear graph.

NAME ______________________ DATE ____________ PERIOD _____

12-2 Reading to Learn Mathematics

Graphing Quadratic Functions

Pre-Activity ***Complete the Mini Lab at the top of page 565 in your textbook. Write your answers below.***

1. Is the relationship between the side length and the area of a square linear or nonlinear? Explain.

2. Describe the shape of the graph.

Reading the Lesson

3. How can you distinguish the equations of quadratic functions from the equations of other functions?

Determine whether each equation represents a quadratic function.

4. $y = 3x - 5$ **5.** $y = 6 - x^2$ **6.** $y = x^3 + 2$ **7.** $y = \frac{8}{x}$

8. Describe the shape of the graph of a quadratic function. What is this shape called?

Determine whether each graph represents a quadratic function.

9.

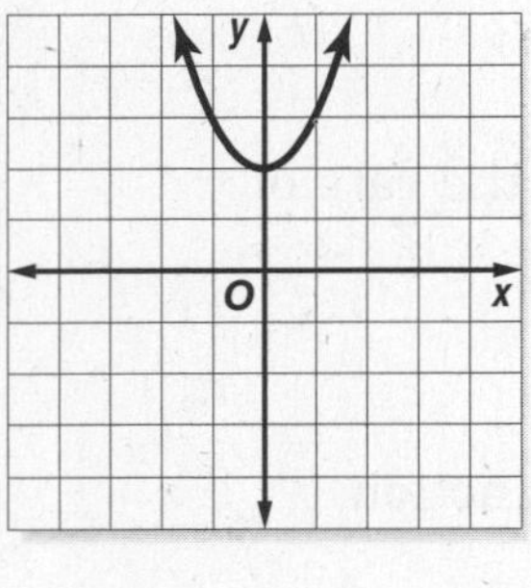

10.

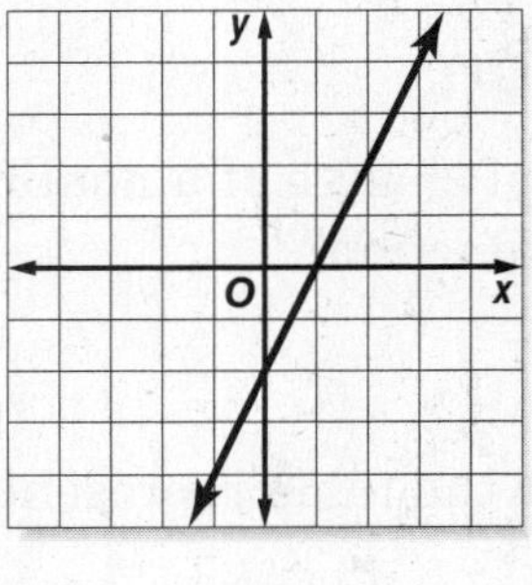

11.

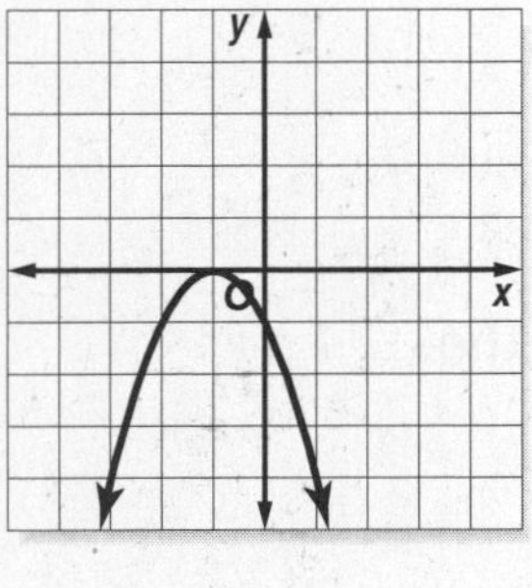

12.

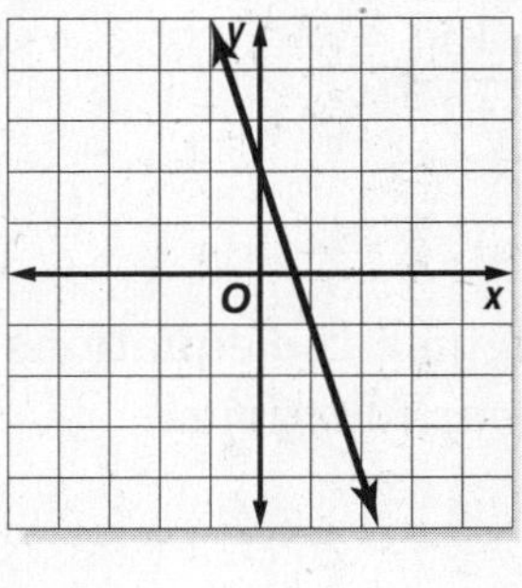

Helping You Remember

13. Work with a partner. Have your partner write a quadratic equation. Then you create a table of values and graph the equation. After a few equations have been completed, switch roles.

Reading to Learn Mathematics

Simplifying Polynomials

Pre-Activity ***Read the introduction at the top of page 570 in your textbook. Write your answers below.***

1. Let q, d, n, and p represent the value of a quarter, a dime, a nickel, and a penny, respectively. Write an expression for the total amount of money in your backpack.

2. Write an expression for the total amount of money in your pocket.

3. Write an expression for the total amount of money in all.

Reading the Lesson

4. What is a monomial?

Determine whether each expression is a monomial. Explain.

5. 8   6. y 7. $3c^2$ 8. $x + 2$

9. Explain what is meant by *like terms*.

Determine whether the expressions are like terms. Explain.

10. d, 6   11. c^3, c 12. $2m^2$, $3m^2$ 13. $2x^2$, $2y^2$

14. Read the Study Tip on page 571 in your textbook. Describe the customary way to write a polynomial in simplified form. What is it called? Which of the polynomials $1 + 2x + x^2$, $2x^3 + 5x^2 + x - 6$, and $6x + x + 3x^3 + 1$ is written in this customary way?

Helping You Remember

15. Work with a partner. Make up a silly statement involving polynomials. For example, *7 apples plus 4 bananas plus 2 apples*. Have your partner write a polynomial to represent the statement and then simplify it. Repeat with your partner making up a silly statement.

NAME ______________________________ DATE ____________ PERIOD _____

12-4

Reading to Learn Mathematics

Adding Polynomials

Pre-Activity ***Complete the Mini Lab at the top of page 574 in your textbook. Write your answers below.***

1. Write the polynomial for the tiles that remain.

2. Use algebra tiles to find $(x^2 + x - 2) + (6x^2 - 5x - 1)$.

Reading the Lesson

3. When adding two polynomials vertically, how should the terms be arranged?

Determine whether each vertical addition can be performed as written. Explain.

4.
$$\begin{array}{r} x^2 + 5x - 6 \\ \underline{(+)\ 2x^2 + 3x + 4} \end{array}$$

5.
$$\begin{array}{r} x^2 + 6 \\ \underline{(+)\ 2x^2 - x} \end{array}$$

Rewrite each sum of polynomials vertically. Then add.

6. $(3d^2 + 14d - 2) + (-d^2 + 3d + 5)$

7. $(2n^2 + 3) + (n^2 - 5n + 1)$

Helping You Remember

8. Given the problem $(3x^2 + 5x - 9) + (x^2 + 2x + 7)$, color code each like term with a different colored pencil. If there are any missing terms, place a box in its space. Then make up your own problem and color code it. Combine the terms with the same color.

NAME ______________________ DATE ____________ PERIOD ____

12-5

Reading to Learn Mathematics

Subtracting Polynomials

Pre-Activity ***Complete the Mini Lab at the top of page 580 in your textbook. Write your answers below.***

1. From the tiles that remain, determine the value of $(x + 4) - (-2x + 3)$.

2. Use algebra tiles to find $(2x^2 + 3x + 5) - (x^2 - x + 2)$.

Reading the Lesson

3. What is the additive inverse of a number or expression?

4. How is the additive inverse related to subtraction of polynomials?

Rewrite each difference of polynomials as a sum of polynomials. Then add.

5. $(7x + 5) - (4x - 3)$

6. $(-3c^2 + 2c - 1) - (-c^2 - c - 2)$

7. $(m^2 + 3m - 6) - (m^2 + 1)$

8. $(-6s + 9) - (4s^2 + 2s - 3)$

Helping You Remember

9. Using a dictionary look up other words containing the prefixes *mono*, *bi*, *and poly*. Give an example of a word containing each prefix and explain how it relates to the lesson.

NAME ______________________________ DATE ____________ PERIOD _____

12-6

Reading to Learn Mathematics

Multiplying and Dividing Monomials

Pre-Activity ***Read the introduction at the top of page 584 in your textbook. Write your answers below.***

1. Examine the exponents of the factors and the exponents of the products in the last column. What do you observe?

Reading the Lesson

2. Under what condition can you multiply two powers by adding their exponents?

Determine whether each multiplication can be performed by adding exponents. Explain.

3. $6^5 \cdot 6^3$

4. $c^2 \cdot c^{10}$

5. $m^3 \cdot n^2$

6. Under what condition can you divide two powers by subtracting their exponents?

Determine whether each division can be performed by subtracting exponents. Explain.

7. $\frac{8^3}{4^2}$

8. $\frac{a^3}{b^3}$

9. $\frac{z^6}{z^5}$

10. Explain how to multiply $(6x^3)(-3x^2)$. Then find the product.

Helping You Remember

11. Explain in your own words the Product of Powers Property and Quotient of Powers Property. Be sure to include an example of each property in your explanation.

NAME ______________________________ DATE ____________ PERIOD _____

12-7 Reading to Learn Mathematics

Multiplying Monomials and Polynomials

Pre-Activity ***Complete the Mini Lab at the top of page 590 in your textbook. Write your answers below.***

1. What is $x(x + 3)$ in simplest form?

Use algebra tiles to find each product.

2. $x(x + 4)$

3. $x(3x + 1)$

4. $2x(x + 3)$

Reading the Lesson

5. Explain how to multiply a polynomial by a monomial.

6. Explain what is done at each step in the following problem.

$3x(x^2 + 5x - 2)$

$= 3x[x^2 + 5x + (-2)]$ ______________________________

$= 3x(x^2) + 3x(5x) + 3x(-2)$ ______________________________

$= 3x^3 + 15x^2 + (-6x)$ ______________________________

$= 3x^3 + 15x^2 - 6x$ ______________________________

7. Give an example from Exercise 6 where the Products of Powers rule is used.

Helping You Remember

8. One way to help you remember to multiply all the terms when multiplying a monomial by a polynomial is to draw arrows from each term outside the parenthesis to each term inside the parenthesis.

$3x(x + 4)$ $(5x - 7)(3)$

Work a problem using arrows and without using arrows. Did you get the same solution? Explain how drawing arrows to the terms is a help.